Chapter 1: The Transformative Potential of AI

Introduction:

The world is on the brink of a remarkable transformation like never before. Artificial Intelligence (AI) has emerged as a force that holds the potential to revolutionize our lives, reshaping the very fabric of society. What was once confined to the realm of science fiction has become a mesmerizing reality, captivating our imagination and challenging our understanding of what it means to be human.

Section 1: Exploring the Marvels of AI

The opening section of this chapter will take you on a journey into the captivating world of AI. We will delve into the history and development of AI, tracing its origins and uncovering the groundbreaking advancements that have brought us to where we are today. We will explore the mind-boggling capabilities of AI, from machine learning to neural networks, and unravel the intricacies of how these technologies have transformed problem-solving and decision-making.

As we navigate through the awe-inspiring landscape of AI, we will reflect upon the immense potential it holds for elevating human cognition. Questions will arise: How can AI augment our ability to process vast amounts of information? Can AI truly surpass human intelligence? What are the ethical implications of AI gaining such power? These inquiries will ignite a sense of wonder and curiosity, propelling us further into the AI revolution.

Section 2: Unraveling a Future of Cooperation

The rise of AI signifies a paradigm shift in the way we perceive competition and cooperation. In this section, we will explore how AI, rather than replacing humans, can work in symbiosis with us to achieve extraordinary feats. We will discover how collaborative efforts combining human ingenuity with AI can unlock innovative solutions to complex problems.

As we delve deeper, we will examine the potential impact of AI on the workforce. The rapid advancement of technology brings forth concerns about job displacement and automation. We will explore the industries and job roles most likely to be affected, and reflect upon the implications this has for our society. What are the ethical and socioeconomic consequences of such transformations? How can we ensure a fair and inclusive future amidst these changes?

Section 3: Pioneering the Unknown: Future Prospects

In the final section, we will cast our gaze toward the horizon and explore the boundless possibilities that lie ahead. We will envision a future where AI and human collaboration have transcended limitations, leading to enhanced creativity, problem-solving, and scientific discoveries. The chapter will touch upon AI's impact on healthcare, transportation, education, and other sectors, sparking conversation and imagination.

As our exploration concludes, we will reflect on the pivotal role of society in shaping the trajectory of AI integration. How can we ensure responsible deployment and use of AI technologies? What are the potential risks and challenges we must collectively address? By contemplating these questions, we will inspire readers to actively participate in the shaping of our AI-integrated future.

Conclusion:

This chapter is only the beginning of a remarkable journey. "The AI-Integrated Human Evolution: From Competition to Cooperation" sets the stage for the exploration of how AI is transforming our lives, amplifying our capabilities, and paving the way toward unprecedented collaboration. Strap in tight as we unlock the immense potential that the fusion of human and artificial intelligence holds. The future is beckoning, and together, we will embark on an unforgettable adventure.

Section 1: Exploring the Marvels of AI (continued)

In this section, we will dive deeper into the captivating world of AI and explore its transformative capabilities. We will examine how AI systems are trained to understand and process complex data, enabling them to perform tasks and make decisions with remarkable accuracy.

Machine learning, a subfield of AI, has revolutionized the way systems process information. Through algorithms and statistical models, machine learning allows AI to learn from large amounts of data, extracting patterns and making predictions or recommendations. Whether it's recognizing objects in images, translating languages, or personalizing recommendations, machine learning has enabled AI to exceed human limitations and achieve astonishing feats.

Neural networks, inspired by the human brain, have emerged as a powerful tool within machine learning. These intricate networks of artificial neurons allow AI systems to learn hierarchical representations of data, enabling them to tackle increasingly complex tasks. Deep learning, a subset of neural networks, has brought breakthroughs in computer vision, natural language processing, and speech recognition, propelling AI into new frontiers.

As AI systems expand their capabilities, they have the potential to augment human thought processes and decision-making. AI algorithms can analyze vast amounts of information in a fraction of the time it would take a human, providing valuable insights and assisting in complex problem-solving. For example, doctors can utilize AI-powered medical imaging systems to detect anomalies and aid in the diagnosis of diseases, significantly improving healthcare outcomes.

However, as AI progresses, questions surrounding the limitations and potential risks arise. Can AI truly surpass human intelligence, or are there fundamental aspects of human cognition that cannot be replicated? The quest to create artificial general intelligence (AGI), with the ability to understand and perform any intellectual task that a human can, remains a captivating but elusive goal. Exploring the boundaries of AI's potential sparks debates on the ethics and implications of unlocking this level of machine intelligence.

Section 2: Unraveling a Future of Cooperation (continued)

As AI continues to advance, a paradigm shift emerges in our perspective on competition and cooperation. Instead of viewing AI as a competitor, we recognize the extraordinary potential of collaborating with these intelligent systems.

Human-AI collaboration has already begun to revolutionize various industries and professions. For example, in the field of design, teams of designers can leverage AI tools to generate innovative ideas and solutions, augmenting their creative processes. Financial analysts can benefit from AI algorithms that process vast amounts of market data, providing valuable insights for investment decisions.

However, the integration of AI into the workforce also raises concerns about job displacement and societal inequalities. It is essential to ensure a fair transition, where AI technologies complement human labor rather than replace it. By focusing on reskilling and upskilling programs, societies can empower individuals to embrace AI and adapt to the changing job landscape. Additionally, policies addressing the ethical, privacy, and transparency challenges posed by AI integration are vital to creating a future where humans and AI thrive collaboratively.

Section 3: Pioneering the Unknown: Future Prospects (continued)

Looking to the future, the potential impact of AI on sectors like healthcare, transportation, education, and more is immense. In healthcare, AI-driven diagnosis and personalized treatment plans have the ability to revolutionize patient care. Self-driving cars, powered by AI algorithms, promise safer and more efficient transportation systems. In education, AI-powered personalized learning platforms can help tailor education to individual needs, enhancing student engagement and outcomes.

The integration of AI technologies into these sectors and countless others opens possibilities for societal advancements and human progress. However, it also requires careful attention to potential risks and challenges. Questions surrounding data privacy, algorithmic biases, and accountability must be addressed to ensure that the benefits of AI are distributed equitably and that no one is left behind.

Conclusion:

Chapter 1 has taken us on a captivating journey through the transformative potential of AI. We have explored the marvels of AI in understanding complex data and augmenting human capabilities. We have also delved into the paradigm shift toward cooperation, highlighting the importance of human-AI collaboration. Looking ahead, we have glimpsed into a future where AI integration revolutionizes various sectors while navigating the ethical and societal challenges that lie ahead.

As we conclude this chapter, let us embrace the awe-inspiring potential of AI while remaining mindful of the ethical considerations and the necessity of shaping its trajectory collectively. The journey ahead is thrilling and full of promise, and together, we can navigate the unexplored territories of human-AI cooperation.

Section 1: Exploring the Marvels of AI (continued)

In this extended section, we will further delve into the captivating world of AI and explore its transformative capabilities across a range of fascinating topics. Buckle up and get ready for a deeper dive into the marvels of artificial intelligence!

1. Reinforcement Learning: Beyond Supervised Learning

While supervised and unsupervised learning are commonly discussed in the context of AI, reinforcement learning presents another exciting approach. Reinforcement learning involves an AI agent interacting with an environment, learning through trial and error to maximize rewards. This approach has been instrumental in creating AI-powered agents that outperform humans in complex games like Go and chess.

2. Natural Language Processing: Opening Up Communication Channels

Natural language processing (NLP) focuses on enabling computers to understand, interpret, and generate human language. With advancements in NLP, chatbots, virtual assistants, and voice-activated systems have become more intuitive and capable of engaging in human-like conversations. From language translation to sentiment analysis and text generation, NLP has become an indispensable tool in our increasingly connected world.

3. Computer Vision: Unveiling the Visual World

Computer vision empowers AI systems to analyze and understand visual data, much like the human visual system. From object recognition and image classification to video analysis and facial recognition, computer vision has countless applications across industries. Cutting-edge AI models, such as convolutional neural networks (CNNs), have achieved remarkable accuracy in tasks like image recognition, enabling AI systems to perceive the world with stunning clarity.

4. Robotics: Bridging the Physical and Virtual Worlds

Integrating AI with robotics has given rise to intelligent machines capable of perceiving their environment and autonomously making decisions. Collaborative robots, or cobots, are designed to work alongside humans, assisting with repetitive or physically demanding tasks. From manufacturing and logistics to healthcare and exploration, the potential for robots equipped with AI algorithms is vast, revolutionizing industries and transforming the way we interact with machines.

5. AI in Healthcare: Revolutionizing Patient Care

The healthcare industry stands to benefit immensely from the integration of AI technologies. AI-powered systems can analyze medical images to detect abnormalities and aid in early diagnosis. Machine learning algorithms can assist in predicting disease progression and optimize treatment plans. Virtual nurses and chatbots can provide 24/7 patient support, answering queries and offering guidance. AI is even helping to develop new drugs and vaccines by rapidly analyzing vast amounts of biomedical data.

6. AI and Climate Change: Addressing Urgent Environmental Challenges

The environmental crisis calls for innovative solutions, and AI is playing a significant role in addressing the challenges associated with climate change. AI-powered systems are being used to optimize energy consumption, predict weather patterns, analyze environmental data, and monitor deforestation. By harnessing AI's capabilities, we can gain valuable insights and develop sustainable strategies to mitigate the impact of climate change on our planet.

7. AI for Social Good: Empowering Communities

AI has the potential to make a positive impact on society by addressing social challenges and promoting social good. For example, AI tools are being employed to improve access to education in underserved areas, enable early detection of diseases in developing regions, and aid disaster response efforts by analyzing satellite imagery and social media data. By harnessing AI ethically and inclusively, we can work toward building a more equitable and just world.

Conclusion:

Chapter 1 has taken us on an expanded journey through the transformative potential of AI, covering a wide range of captivating topics. From reinforcement learning and natural language processing to computer vision, robotics, and AI in healthcare and climate change, we have explored the diverse applications of AI in various domains. Additionally, we have glimpsed AI's role in empowering communities and achieving social good.

As we conclude this chapter, let us acknowledge the limitless possibilities that lie ahead as AI continues to advance and reshape our world. However, it is crucial to remain vigilant about the ethical considerations, transparency, and inclusivity in the development and implementation of AI technologies. By fostering a collaborative and responsible approach, we can unlock the full potential of AI while ensuring it serves the best interests of humanity.

Section 2: Unveiling Advanced AI Technologies

In this expanded section, we will delve even deeper into the exciting advancements within the field of AI, exploring cutting-edge technologies and their potential impact on various industries. Let's dive right in!

1. Generative Adversarial Networks (GANs): Fueling Creative AI

Generative Adversarial Networks (GANs) have revolutionized the field of AI by enabling machines to create new and realistic content. GANs consist of two competing neural networks—the generator and the discriminator—working against each other. This adversarial setup allows GANs to generate entirely new images, videos, and even text-based content that closely resembles human-created content. From generating realistic faces to creating digital art and designing virtual worlds, GANs are pushing the boundaries of AI creativity.

2. Explainable AI: Understanding the Decision-Making Process

As AI systems become increasingly complex, understanding how decisions are made within these systems becomes crucial. Explainable AI focuses on developing methods and techniques that provide insights into the decision-making process of AI models and algorithms. By explaining the reasoning behind AI-based recommendations and predictions, explainable AI helps build trust, enhance transparency, and ensure that AI is fair and ethical.

3. Edge Computing: Bringing AI Closer to the Action

Edge computing involves processing and analyzing data closer to the source rather than relying on centralized cloud infrastructure. This approach is particularly useful in AI applications requiring real-time or low-latency responses, such as autonomous vehicles or Internet of Things (IoT) devices. By executing AI algorithms at the edge, data can be processed more efficiently, reducing bandwidth usage and improving response times.

4. Quantum Computing: Unlocking New Frontiers

Quantum computing holds tremendous promise in its ability to exponentially increase computational power. With this newfound power, AI algorithms can be drastically enhanced to solve even more complex problems. Quantum machine learning and quantum neural networks are emerging areas, blending the power of quantum computing with AI to unlock new frontiers in problem-solving, optimization, and data analysis.

5. AI in Finance: Revolutionizing the Financial Industry

The financial industry has embraced AI's capabilities to optimize decision-making, fraud detection, risk assessment, and trading strategies. AI-powered algorithms can analyze vast amounts of financial data, identify patterns, and make accurate predictions. Robo-advisors have also emerged, providing personalized investment advice based on individual financial goals. With AI's speed and accuracy, financial institutions can streamline operations, improve customer experiences, and enhance overall efficiency.

6. AI in Entertainment: Enhancing Creativity and Immersion

From personalized recommendations on streaming platforms to AI-generated music and virtual reality experiences, AI is transforming the entertainment industry. Recommendation systems powered by AI algorithms analyze individual preferences and behavior to suggest relevant movies, shows, and songs. AI-driven virtual reality (VR) and augmented reality (AR) technologies bring immersive experiences to life, revolutionizing gaming, storytelling, and interactive media.

7. AI in Agriculture: Nurturing Sustainable Food Production

The agriculture industry faces the challenge of feeding a growing global population while minimizing environmental impact. AI technologies offer solutions by optimizing resource management, crop yield prediction, pest detection, and precision agriculture. Drones equipped with AI-powered cameras can monitor and analyze crops, identifying potential issues early on. AI algorithms can also process data from sensors to regulate water and fertilizer usage, reducing waste and improving sustainability.

Conclusion:

With the expanded exploration of advanced AI technologies, we have uncovered an array of cutting-edge research and applications that push the boundaries of what AI can achieve. From GANs and explainable AI to edge computing, quantum computing, and AI's impact on finance, entertainment, and agriculture, we see AI's transformative potential across diverse industries.

As we move into the next chapter, we will continue to unravel the complexities and advancements within the AI field. Remember, the journey of AI is an ever-evolving one, and there are always new breakthroughs and exciting discoveries around the corner.

8. AI in Healthcare: Transforming Patient Care

The healthcare industry is experiencing a remarkable transformation with the integration of AI technologies. AI-assisted diagnosis systems can analyze medical images, such as X-rays and MRIs, to provide accurate and timely assessments, aiding in the early detection of diseases. AI also plays a vital role in personalized medicine, utilizing patient data and genetic information to tailor treatment plans and predict patient outcomes. Additionally, AI-powered chatbots and virtual assistants help streamline administrative tasks and provide 24/7 patient support, enhancing overall healthcare accessibility and efficiency.

9. AI Ethics: Ensuring Human-Centric and Responsible AI

As AI technologies become more sophisticated and pervasive, considering the ethical implications and ensuring responsible AI development is of utmost importance. The field of AI ethics focuses on identifying and addressing potential biases, privacy concerns, fairness issues, and the risks associated with autonomous systems. By incorporating ethical guidelines, transparency, and accountability into AI development, we can ensure that AI systems are designed to benefit humanity and respect fundamental values and rights.

10. AI in Education: Personalized Learning and Skill Development

AI is reshaping the education landscape by enabling personalized learning experiences tailored to each student's unique needs and abilities. With AI-powered adaptive learning platforms, educational institutions can analyze individual student performance data and provide targeted recommendations and interventions. Virtual tutors and intelligent educational assistants simulate human teachers, answering questions and assisting students in their learning journey. AI also plays a vital role in skills assessment, helping individuals acquire the necessary competencies for the jobs of the future.

11. AI in Transportation: Revolutionizing Mobility

From self-driving cars to smart traffic management systems, AI is revolutionizing transportation and mobility. Autonomous vehicles powered by AI algorithms have the potential to enhance road safety, reduce accidents, and transform the way we commute. AI-powered predictive analytics can optimize traffic flow, minimize congestion, and improve urban planning. Additionally, AI algorithms aid logistics and supply chain management, optimizing routes, and reducing transportation costs.

12. AI in Climate Change Mitigation: Sustainable Solutions

Addressing climate change requires innovative solutions, and AI offers valuable contributions in this area. AI-powered climate models can analyze vast amounts of environmental data, improving weather predictions and facilitating early warning systems for natural disasters. Energy optimization using AI algorithms can help reduce carbon footprints by optimizing energy consumption and promoting renewable energy sources. AI can also assist in wildlife preservation by analyzing satellite data and monitoring endangered species' habitats.

13. AI in Cybersecurity: Safeguarding Digital Infrastructures

As the digital landscape grows, so does the importance of cybersecurity. AI technologies are playing a crucial role in detecting and preventing cyber threats. AI-powered systems can analyze network traffic patterns and identify anomalies indicative of potential attacks. Machine learning algorithms can continuously learn and adapt, improving the identification and mitigation of emerging threats. AI-based authentication and biometric technologies provide enhanced security measures, protecting sensitive data from unauthorized access.

As we continue exploring these fascinating applications of AI, we are bound to uncover more intriguing possibilities. The rapidly evolving field of AI offers significant potential for positive change across various sectors. If there's any specific area you'd like to delve deeper into or any other aspect you'd like me to explore, please let me know!

14. AI in Finance: Enhancing Financial Services

AI is reshaping the financial industry by automating routine tasks, improving fraud detection, and providing more accurate risk assessments. Chatbots and virtual assistants powered by AI can handle customer inquiries and provide personalized financial advice. AI algorithms analyze vast amounts of financial data in real-time, enabling faster and more informed investment decisions. Additionally, AI-driven algorithms can detect suspicious patterns and behaviors, assisting in fraud prevention and ensuring the security of financial transactions.

15. AI in Entertainment: Enriching User Experiences

AI technologies have significantly impacted the entertainment industry, enabling personalized recommendations, content creation, and immersive experiences. Streaming platforms and music services use AI algorithms to analyze user preferences and behavior to suggest relevant movies, series, or songs. AI-powered algorithms can generate realistic visuals, enhance special effects, and streamline post-production processes in the film and gaming industries. Virtual reality (VR) and augmented reality (AR) experiences are also enhanced by AI, providing interactive and immersive virtual worlds.

16. AI and Social Media: Transforming Digital Interactions

AI has revolutionized social media platforms by enhancing user experiences, content moderation, and targeted advertising. AI algorithms analyze user data, such as preferences, browsing history, and engagement patterns, to customize content recommendations and targeted advertisements. AI-powered content moderation systems help flag and remove inappropriate or harmful content, ensuring a safer online environment. Additionally, AI-driven sentiment analysis tools analyze social media posts and comments to gauge public opinion and monitor trends.

17. AI in Robotics: Advancing Automation and Services

Robotics and AI technologies go hand in hand, powering autonomous robots and intelligent machines. Robots equipped with AI algorithms can perform complex tasks, from industrial assembly line operations to household chores. AI-powered robotic assistants provide support in healthcare settings, assisting patients with daily activities and monitoring vital signs. Drones with AI capabilities enable efficient package delivery and assist in disaster management and search-and-rescue missions. As AI continues to advance, the potential for robotics in various industries is ever-expanding.

18. AI in Agriculture: Enhancing Efficiency and Sustainability

AI is revolutionizing agriculture by optimizing crop yields, reducing waste, and promoting sustainable practices. AI-powered systems analyze soil and weather data to provide accurate recommendations for irrigation, fertilization, and pest control. Drones equipped with AI algorithms facilitate crop monitoring, helping identify areas needing attention, and enabling targeted interventions. AI-driven robotics streamlines harvesting and sorting processes, reducing labor costs and improving efficiency. These advancements contribute to more sustainable and productive agricultural practices.

19. AI in Retail: Reinventing the Shopping Experience

AI technologies have transformed the retail landscape by personalizing customer experiences, improving inventory management, and optimizing pricing strategies. AI-powered chatbots and virtual shopping assistants guide customers through their purchasing journey, providing product recommendations and answering inquiries. AI algorithms analyze customer behavior and preferences to predict demand, ensuring optimal stock levels and reducing inventory costs. Additionally, AI-driven dynamic pricing systems adjust product prices in real-time, reflecting market trends and maximizing profit margins.

20. AI in Space Exploration: Expanding the Horizons of Science

AI is playing a crucial role in space exploration, assisting in mission planning, data analysis, and robotics. AI algorithms help analyze vast amounts of astronomical data, aiding in the discovery of exoplanets, gravitational waves, and cosmic events. Robotic systems equipped with AI capabilities are used to explore extraterrestrial environments autonomously. AI also assists in spacecraft navigation and control during complex missions, ensuring precise maneuvering and successful exploration.

As AI continues to advance, its impact on various industries and aspects of our lives is bound to grow even further. If there's a particular topic or application you'd like to explore in more detail, feel free to let me know!

21. AI in Healthcare: Transforming Medical Diagnosis and Treatment

AI is revolutionizing healthcare by improving medical diagnosis, treatment planning, and patient monitoring. AI algorithms analyze patient data, such as medical records, imaging scans, and genetic profiles, to assist in the early detection and diagnosis of diseases. Machine learning models help identify patterns and predict treatment outcomes, helping physicians make informed decisions. AI-powered robotic surgery systems enable precise and minimally invasive procedures, reducing patient recovery time. AI technologies also have the potential to streamline healthcare operations, such as scheduling appointments, optimizing resource allocation, and improving patient experience.

22. AI in Transportation: Enabling Autonomous Vehicles and Smart Traffic Systems

AI is driving advancements in transportation, with a focus on autonomous vehicles and improving traffic management systems. Self-driving cars powered by AI algorithms incorporate sensors, cameras, and machine learning models to perceive and react to the environment. AI algorithms help optimize route planning, reducing congestion and improving fuel efficiency. Additionally, AI-driven traffic control systems analyze real-time data to optimize traffic flow, reducing travel times and minimizing accidents. The transportation industry is experiencing a transformative shift with the integration of AI technologies.

23. AI in Cybersecurity: Enhancing Threat Detection and Prevention

AI plays a vital role in strengthening cybersecurity defenses against evolving threats. AI algorithms can analyze large volumes of data and detect patterns associated with malicious activities. They help identify anomalies, detect potential security breaches, and alert security teams for immediate action. AI-driven security solutions can autonomously respond to threats, mitigating risks and reducing response times. By continuously learning from new threats, AI enables proactive protection, keeping sensitive data and systems secure.

24. AI for Personalized Education: Tailoring Learning Experiences

AI technologies are revolutionizing education by personalizing the learning experience and adapting to individual needs. AI-powered educational platforms use adaptive algorithms to analyze student performance, identify strengths and weaknesses, and provide tailored learning materials. Intelligent tutoring systems engage students, provide real-time feedback, and offer targeted recommendations for improvement. AI chatbots assist students in their inquiries and provide support outside the classroom. With AI, education becomes more accessible and flexible, catering to diverse learning styles and needs.

25. AI in Energy: Optimizing Resource Management and Sustainability

AI is being utilized in the energy sector to improve resource management, enhance efficiency, and promote sustainability. AI algorithms analyze energy consumption data to optimize resource allocation, reduce waste, and improve energy efficiency. Predictive analytics models help forecast energy demand, aiding in efficient production planning and distribution. AI-driven systems also enhance renewable energy integration by optimizing output and storage solutions. By leveraging AI technologies, the energy sector can transition to cleaner, more sustainable practices.

The applications of AI are varied and constantly expanding, offering significant benefits across numerous industries. If there's a specific area you'd like to explore further or if you have any questions, feel free to let me know!

26. AI in Agriculture: Improving Crop Yield and Efficient Farming

AI is revolutionizing agriculture by enhancing crop yield, reducing resource waste, and enabling efficient farming practices. AI technologies analyze data from sensors, satellites, and drones to monitor soil conditions, crop health, and weather patterns. Machine learning models help farmers optimize irrigation schedules, detect diseases and pests early, and determine optimal harvesting times. AI-powered robots and drones assist in tasks such as planting, harvesting, and monitoring crops. By leveraging AI, farmers can improve productivity, reduce environmental impact, and ensure sustainable agriculture practices.

27. AI in Retail: Enhancing Customer Experience and Personalizing Recommendations

AI is transforming the retail industry by providing personalized shopping experiences, optimizing inventory management, and improving customer service. AI algorithms analyze customer data, such as purchase history, browsing behavior, and social media interactions, to deliver personalized product recommendations. AI-powered chatbots and virtual assistants offer real-time support to customers, handling inquiries and providing assistance. AI technologies also help retailers optimize supply chain operations, predict demand, and prevent stockouts. By leveraging AI, retailers can create seamless and tailored experiences for their customers.

28. AI in Finance: Improving Fraud Detection and Enhancing Investment Strategies

AI is reshaping the finance industry by enhancing fraud detection, automating processes, and improving investment strategies. AI algorithms analyze vast amounts of financial data to detect patterns associated with fraudulent activities and flag suspicious transactions. Machine learning models assist in credit scoring, loan underwriting, and risk management. AI-powered chatbots and virtual assistants provide personalized financial advice and support to customers. AI technologies also enable algorithmic trading, optimizing investment decisions based on vast amounts of market data. With AI, the finance industry becomes more efficient, secure, and accessible.

29. AI in Entertainment: Transforming Content Creation and Personalized Recommendations

AI technologies are revolutionizing the entertainment industry by transforming content creation and enhancing personalized recommendations for users. AI algorithms analyze user preferences, viewing history, and feedback to deliver personalized content recommendations on streaming platforms. AI-powered content creation tools assist in video editing, special effects, and music production. Voice recognition and natural language processing enable more sophisticated human-computer interactions, enhancing virtual assistants and voice-controlled devices. By leveraging AI, the entertainment industry can deliver engaging and tailored experiences to audiences.

30. AI in Social Media: Improving User Experience and Content Moderation

AI plays a crucial role in improving user experience and content moderation on social media platforms. AI algorithms analyze user behavior, preferences, and interactions to provide a personalized feed of content. Sentiment analysis and natural language processing help identify and filter out inappropriate or harmful content. AI technologies also assist in detecting fake accounts, spam, and cybersecurity threats. By leveraging AI, social media platforms can create safer and more engaging environments for users worldwide.

31. AI in Healthcare: Revolutionizing Diagnosis and Treatment

AI is transforming healthcare by enabling faster and more accurate diagnosis, personalized treatment plans, and drug discovery. Machine learning algorithms analyze medical images, such as X-rays and MRIs, to detect abnormalities and assist in early disease detection. AI-powered chatbots and virtual assistants provide patients with symptom analysis and medical advice. Natural language processing helps analyze medical literature to suggest potential treatment options. AI technologies also assist in predicting patient outcomes, optimizing hospital operations, and improving patient care.

32. AI in Transportation: Enhancing Safety and Efficiency

AI is revolutionizing the transportation industry by improving safety, optimizing routes, and facilitating autonomous vehicles. AI algorithms analyze real-time traffic data, weather conditions, and historical patterns to optimize traffic flow and reduce congestion. Computer vision and sensor fusion enable autonomous vehicles to perceive and navigate the environment. AI technologies also enhance driver assistance systems, providing features like lane-keeping, adaptive cruise control, and collision avoidance. By leveraging AI, transportation becomes safer, more efficient, and sustainable.

33. AI in Energy: Optimizing Resource Management and Grid Efficiency

AI plays a crucial role in the energy sector by optimizing resource management, predicting demand, and improving grid efficiency. AI algorithms analyze historical energy consumption data to predict future demand patterns, allowing energy providers to adjust production and distribution accordingly. Machine learning models help optimize renewable energy generation and storage systems. AI technologies also assist in monitoring and maintaining the reliability of power grids, detecting anomalies and predicting failures. By leveraging AI, the energy sector can achieve more efficient and sustainable operations.

34. AI in Education: Personalizing Learning and Enhancing Educational Tools

AI is reshaping the education sector by personalizing learning experiences, improving educational tools, and enabling adaptive tutoring systems. AI-powered algorithms analyze student performance data to identify individual learning styles and provide personalized recommendations and feedback. Natural language processing helps develop intelligent tutoring systems that assist students with their assignments and provide instant feedback. AI technologies also assist in automating administrative tasks, like grading assessments and managing schedules. By leveraging AI, education becomes more accessible, engaging, and effective.

35. AI in Gaming: Enhancing Immersion and Adaptive Gameplay

AI technologies have made significant advancements in the gaming industry, enhancing player experiences and enabling adaptive gameplay. AI algorithms analyze player behavior to create intelligent non-player characters (NPCs) that respond dynamically to each player's actions. AI-powered procedural generation generates unique and realistic game environments, reducing manual development efforts. Machine learning models improve game physics, AI opponents, and adaptive difficulty levels. By leveraging AI, gaming becomes more immersive, challenging, and personalized.

Chapter 2: Unleashing the Power of Collective Intelligence

In a world that often celebrates individual achievements, it's time to explore the astounding potential of cooperation and collective intelligence. Prepare yourself for an exhilarating journey as we unveil groundbreaking research, share inspiring stories, and challenge the conventional notions of success. Together, let's redefine the very essence of achievement and discover the tremendous power that lies within our shared destiny.

Section 1: The Rise of Collaborative Problem Solving

Reflecting upon history, we find numerous instances where the accomplishments of individuals were not born solely out of their own brilliance, but rather through collaboration and shared wisdom. From the ancient Egyptian pyramids to the technological wonders of today, the power of collective effort has consistently sparked human progress.

Let us delve into a future where collaboration becomes the norm rather than the exception. Imagine scientists from diverse backgrounds coming together to tackle humanity's most pressing challenges. Through their combined knowledge and expertise, groundbreaking solutions emerge, leading to breakthroughs in medicine, climate change mitigation, and beyond. The potential for innovation and problem-solving is limitless when we pool our collective intelligence.

Section 2: The Dynamics of Collective Intelligence

Now, dear reader, picture a world where technology facilitates seamless collaboration among individuals, transcending geographical boundaries and time constraints. Virtual platforms and advanced communication tools enable people across the globe to connect effortlessly, sharing their unique perspectives and building upon each other's ideas.

But let us not forget the challenges and risks that accompany this exciting frontier. As we embrace the power of collective intelligence, it is crucial to navigate the potential pitfalls. We must address concerns surrounding intellectual property, privacy, and security, ensuring that the collective effort does not succumb to exploitation or misuse. Additionally, we must remain vigilant about fostering an inclusive and diverse collaborative environment, ensuring that all voices are heard and valued.

Section 3: From Competition to Cooperation

In a world that often glorifies cutthroat competition, we invite you to envision a future where cooperation supersedes rivalry. Imagine a shift in societal values where collective achievements are celebrated, and individual successes are seen as contributions to a larger tapestry of progress.

Education, a cornerstone of shaping minds, evolves to prioritize teamwork, empathy, and collaboration. Students engage in interactive group projects, cultivating essential skills for the complex challenges of the future. Society recognizes that the breakthroughs that define our era will arise not from isolated geniuses but from diverse teams united by a common purpose.

Section 4: The Uncharted Frontiers of Collective Potential

As we embark on this transformative journey, the possibilities for collective intelligence are vast and unprecedented. Imagine communities collaborating to tackle climate change, leveraging their local knowledge and global expertise to implement sustainable solutions. Picture a world where grassroots movements, driven by shared values, topple oppressive structures and foster greater inclusivity.

While we embrace the boundless potential of collective intelligence, we must acknowledge the need for responsible stewardship. Ethical considerations become paramount, shaping our approach to decision-making, data privacy, and the allocation of resources.

Conclusion: Our Shared Destiny Awaits

Dear reader, as we conclude this visionary exploration of collective intelligence, remember that the power to shape our destiny lies within each of us. By embracing collaboration and synergy, we can transcend our individual limitations and achieve unprecedented heights.

Let us step into this future hand in hand, recognizing that true greatness is found not in solitary pursuits, but in the shared journey towards a brighter world. Together, we have the power to create a future where collective intelligence, guided by ethics and inclusion, leads us towards a dazzling horizon of possibilities.

Are you ready to join the ranks of those who dare to dream and collaborate for a better tomorrow?

Section 5: Nurturing a Collaborative Ecosystem

To fully harness the power of collective intelligence, we need to cultivate a vibrant ecosystem that encourages collaboration, information sharing, and interconnectivity. Institutions, organizations, and governments play a crucial role in providing the platforms and resources necessary for collective efforts to thrive.

Imagine a future where companies shift their focus from cutthroat competition to cooperative partnerships. Industries join forces, pooling their resources, expertise, and innovative ideas to solve complex challenges. By sharing knowledge and collaborating, these organizations create a rising tide that lifts all boats, pushing the boundaries of progress even further.

Furthermore, governments recognize the potential of collective intelligence and invest in initiatives that foster collaboration and innovation. They create open data policies, encourage collaboration between researchers and policymakers, and establish funding programs to support collective problem-solving efforts. By providing the infrastructure and support necessary for collaboration, governments become catalysts for a vibrant ecosystem of collective intelligence.

Section 6: The Role of Artificial Intelligence

In our quest to unravel the power of collective intelligence, we cannot overlook the role of artificial intelligence (AI). AI technologies, with their ability to analyze vast amounts of data and uncover patterns, have the potential to augment human collaborative efforts.

Imagine AI algorithms that can identify complementary skill sets and bring together individuals with diverse expertise to tackle complex projects. These algorithms would connect like-minded thinkers and foster an environment where collaboration blossoms. Additionally, AI can assist in gathering and synthesizing information from various sources, enabling collective decision-making processes that leverage the wisdom of the crowd.

Nevertheless, we must tread carefully as we integrate AI into our collaborative endeavors. Ensuring transparency, accountability, and ethical guidelines in AI development and deployment becomes paramount. By addressing potential biases, privacy concerns, and the preservation of human agency, we can strike a balance where AI enhances our collective intelligence while respecting our shared values.

Section 7: Overcoming Barriers to Collaboration

As we embark on this transformative journey toward collective intelligence, we need to address the barriers that hinder collaboration. Deep-seated cultural norms that emphasize individualism, competition, and ego must be confronted and replaced with a culture of cooperation and shared success.

Education plays a pivotal role in instilling the collaborative mindset early on. By incorporating teamwork, communication, empathy, and conflict resolution skills into curricula, we equip future generations with the tools needed to thrive in a collaborative world. Likewise, organizations and communities can create spaces and platforms that foster trust, respect, and equal participation, breaking down barriers and encouraging collaboration across diverse backgrounds.

Section 8: Challenges and Risks on the Horizon

While the promises of collective intelligence are alluring, we must remain cognizant of the challenges and risks that accompany this paradigm shift. As we rely more heavily on collective decision-making processes, the possibility of groupthink and the diffusion of responsibility may arise. It becomes crucial to establish frameworks for critical thinking, ensuring that the quality and diversity of contributions are maintained.

Additionally, issues of security, trust, and data integrity emerge in collaborative environments. Protecting sensitive information, preventing malicious exploitation, and maintaining the privacy of participants become key considerations in this interconnected landscape. Ethical guidelines and robust cybersecurity measures must be developed and continually refined to address these challenges.

Conclusion: Pioneering a New Frontier

Dear reader, as we conclude this chapter on transcending individualism and embracing collective intelligence, let us reflect on the immense possibilities that lie before us. We have explored a future where collaboration, guided by ethics, inclusion, and technological innovation, reshapes the very fabric of human achievement.

By nurturing a collaborative ecosystem, leveraging the power of AI, and overcoming barriers to collaboration, we can unlock the true potential of collective intelligence. Let us courageously embrace this new frontier, recognizing that our shared destiny calls upon us to rise above individual aspirations and collaborate for the betterment of all.

Are you ready, dear reader, to embark on this transformative journey, where the cooperative pursuit of greatness leads us toward a future that surpasses our wildest imagination? Together, let us pioneer this new era of collective intelligence and forge a path to a brighter and more harmonious tomorrow.

Chapter 3: Applications of Collective Intelligence

Section 1: Problem Solving and Decision Making

One of the most evident applications of collective intelligence is in problem solving and decision making. When tackling complex challenges, the collective knowledge and perspectives of a diverse group can lead to more innovative and effective solutions. By tapping into the wisdom of the crowd, we can leverage a broader range of expertise, insights, and experiences, ultimately arriving at better-informed decisions.

In this context, crowdsourcing platforms have emerged as powerful tools for harnessing collective intelligence. These platforms enable individuals from diverse backgrounds and locations to contribute their ideas and expertise to solve specific problems. Whether it's seeking innovative solutions for scientific research, social issues, or business challenges, crowdsourcing can leverage the power of collective problem-solving on a global scale.

Section 2: Prediction and Forecasting

Collective intelligence also finds extensive applications in prediction and forecasting. By aggregating the predictions or opinions of a large group, we can obtain more accurate and reliable forecasts than relying on single experts. This concept, known as the "wisdom of the crowd," capitalizes on the idea that while individual judgments may be flawed or biased, the collective average tends to be more accurate.

Prediction markets, for example, tap into collective intelligence by allowing individuals to buy and sell contracts based on their beliefs about future events. The market prices of these contracts reflect the aggregated opinions and predictions of the participants, providing valuable insights and predictions on various topics, from economic trends to election outcomes.

Section 3: Citizen Science

Collective intelligence has also found a home in the realm of citizen science. With the proliferation of technology and connectivity, ordinary individuals can now actively contribute to scientific research and data collection. Citizen science projects engage volunteers in tasks such as data collection, classification, and analysis, allowing them to contribute their observations and insights to scientific studies.

From birdwatching to monitoring environmental changes, citizen science projects have contributed significantly to various research fields. By harnessing the collective power of passionate individuals around the world, scientists can access vast amounts of data and uncover patterns that would otherwise be challenging to obtain.

Section 4: Collaborative Creation

Another exciting application of collective intelligence is in collaborative creation. Online platforms and communities have enabled individuals to come together and collaboratively create content, art, software, and more. Wikipedia, a prime example of collaborative creation, has revolutionized the way knowledge is shared and collectively built by allowing anyone to contribute and edit articles.

Open-source software development is another field where collective intelligence plays a vital role. Developers worldwide collaborate on projects, sharing code, contributing improvements, and collectively enhancing the software. This participatory approach has led to the creation of high-quality, freely available software that benefits users globally.

Section 5: Social and Political Engagement

Collective intelligence is also transforming social and political engagement. Online platforms and social media empower individuals to voice their opinions, share ideas, and collectively mobilize for causes they believe in. Movements, such as the Arab Spring and various social justice initiatives, have demonstrated the potential for collective action facilitated by digital technologies.

In addition to political engagement, collective intelligence can also enhance public decision-making processes. Participatory democracy and deliberative platforms enable citizens to contribute their insights and perspectives when making policy decisions and shaping the future of their communities. By involving a diverse range of voices, these initiatives can lead to more inclusive and representative outcomes.

Conclusion: Unlocking the Power of Collective Intelligence

Dear reader, as we conclude this chapter on the applications of collective intelligence, we have witnessed the breadth and depth of its transformative potential. From problem solving and decision making to citizen science, collaborative creation, and social and political engagement, collective intelligence is reshaping how we approach, tackle, and solve complex problems.

As technological advancements continue to spark new possibilities, the power of collective intelligence will only proliferate. By embracing collaboration, fostering inclusive environments, and harnessing the collective wisdom of diverse individuals, we can unlock solutions and innovations that surpass what any one individual or organization could achieve alone.

So, dear reader, let us continue on this journey of exploring and harnessing the power of collective intelligence. In the chapters ahead, we will delve deeper into the tools, strategies, and frameworks that can help us maximize the potential of collaborative efforts. Together, we can pioneer a future where collective intelligence becomes a cornerstone of human progress.

Chapter 4: The Science Behind Collective Intelligence

Section 1: Understanding Collective Intelligence

In order to fully grasp the power and potential of collective intelligence, it is important to delve into the science behind it. Collective intelligence refers to the capacity of groups or collectives to exhibit enhanced problem-solving, decision-making, and creative abilities that surpass those of individual members. This phenomenon arises from the interaction and collaboration among individuals, resulting in emergent properties that are not present at the individual level.

The concept of collective intelligence draws from various disciplines, including social psychology, cognitive science, network theory, and computer science. By understanding the underlying mechanisms and dynamics, we can develop strategies and tools to harness collective intelligence effectively.

Section 2: Factors Influencing Collective Intelligence

Several factors influence the effectiveness of collective intelligence within a group. Let's explore some of the key factors:

1. Diversity: Diversity, both in terms of knowledge and perspectives, is crucial for unlocking the full potential of collective intelligence. When individuals bring different insights and expertise to the table, it can lead to more innovative and creative solutions. Diversity expands the range of possible ideas and viewpoints, promoting effective problem-solving.

2. Cognitive Abilities: The cognitive abilities of group members also play a significant role in collective intelligence. A group composed of individuals with diverse cognitive strengths, such as analytical thinking, creativity, and critical reasoning, can leverage these skills to tackle different aspects of a problem effectively.

3. Communication and Collaboration: Effective communication and collaboration are essential for harnessing collective intelligence. Open and inclusive communication channels facilitate the sharing of ideas, constructive feedback, and the integration of different perspectives. Collaboration tools and techniques can help foster a collaborative environment where collective intelligence can thrive.

4. Trust and Psychological Safety: Trust and psychological safety within a group create an environment where individuals feel comfortable sharing their ideas and taking risks. When members trust that their contributions will be valued and respected, they are more likely to actively participate and contribute to collective intelligence processes.

Section 3: The Role of Technology

Technological advancements have played a significant role in facilitating and enhancing collective intelligence. Online platforms and tools enable individuals to connect, collaborate, and share knowledge on a global scale. Here are some ways technology supports collective intelligence:

1. Crowdsourcing Platforms: Crowdsourcing platforms allow individuals to contribute their ideas, expertise, and insights to tackle complex problems collectively. These platforms provide a structured framework for collaboration and enable diverse participants to contribute regardless of their location.

2. Social Media: Social media platforms provide a space for individuals to share their perspectives, engage in discussions, and mobilize for collective action. Social media can amplify the voices of diverse individuals, facilitating the exchange of information and ideas at a rapid pace.

3. Collaborative Tools and Software: Various collaborative tools and software facilitate group work and enable simultaneous contributions. These tools range from document editing platforms like Google Docs to project management software like Trello, enhancing communication, coordination, and knowledge sharing within teams.

Section 4: The Wisdom of the Crowd

One of the central concepts underlying collective intelligence is the "wisdom of the crowd." This phenomenon suggests that the aggregated opinions or decisions of a large group tend to be more accurate and reliable than those of individuals. The wisdom of the crowd relies on the principle that individual biases and errors tend to cancel each other out, resulting in a more accurate average or consensus.

Several conditions are necessary to realize the wisdom of the crowd:

1. Independence: Individual judgments should be independent of one another, free from direct influence or pressure. When individuals are influenced by others' opinions or manipulated, the wisdom of the crowd may be compromised.

2. Diversity: A diverse crowd, encompassing a wide range of perspectives, knowledge, and expertise, enhances the wisdom of the crowd. Different viewpoints and insights contribute to more robust collective decisions.

3. Aggregation Mechanisms: Appropriate aggregation mechanisms are needed to synthesize the collective judgments or opinions. These mechanisms can include voting systems, prediction markets, or statistical methods to extract valuable insights from the crowd.

Section 5: Ethical Considerations

While collective intelligence offers tremendous potential, it also raises ethical considerations that must be addressed. Here are some key aspects to consider:

1. Privacy and Data Protection: As collective intelligence often relies on data sharing and collaboration, ensuring the privacy and security of participants' data is paramount. Guidelines and regulations need to be in place to protect individuals' privacy and prevent the misuse of data.

2. Inclusivity and Representation: It is important to ensure that collective intelligence processes include diverse individuals from various backgrounds and perspectives. Exclusion or underrepresentation of certain groups can result in biased outcomes and limit the effectiveness of collective intelligence.

3. Transparency and Accountability: Transparency and accountability are crucial in collective intelligence processes. Clear guidelines, feedback mechanisms, and decision-making processes should be in place to ensure that participants understand how their contributions are being used and how decisions are made.

Conclusion: Harnessing the Power of Collective Intelligence

As we conclude this chapter on the science behind collective intelligence, we gain deeper insights into the mechanisms and dynamics that make collective intelligence effective. Understanding the factors that influence collective intelligence, the role of technology, the wisdom of the crowd, and the ethical considerations involved allows us to leverage collective intelligence in a responsible and effective manner.

By embracing diversity, fostering effective communication and collaboration, and leveraging technological advancements, we can unlock the full potential of collective intelligence. This power has the capacity to revolutionize problem-solving, decision-making, creativity, and social engagement, leading to a brighter and more innovative future.

In the chapters ahead, we will explore practical strategies, tools, and examples that illustrate how collective intelligence can be applied in various domains and contexts. Together, let us continue this journey of harnessing collective intelligence and shape a future where collaboration and collective wisdom drive progress and innovation.

Chapter 5: The Rise of AI and Human Collaboration

In this chapter, we will explore the remarkable symbiotic relationship between humans and artificial intelligence (AI) and how it has revolutionized various fields. AI technologies have enhanced human capabilities, leading to unprecedented collaboration and cooperation.

Advances in AI have paved the way for groundbreaking applications in healthcare, where AI algorithms analyze large amounts of medical data to assist in diagnosis, treatment planning, and drug discovery. By working together with AI, healthcare professionals can make more accurate and timely decisions, ultimately improving patient outcomes.

In the realm of space exploration, AI-powered robots have become indispensable partners. They assist astronauts in complex tasks, such as collecting samples from distant planets or repairing spacecraft. This collaboration between humans and AI has expanded the boundaries of our exploration and deepened our understanding of the universe.

Beyond healthcare and space exploration, AI has also transformed industries like finance, manufacturing, and transportation. Machine learning algorithms have optimized trading strategies, automated production lines, and enhanced safety in autonomous vehicles. These collaborations have increased efficiency and productivity while freeing humans to focus on more creative and strategic tasks.

The rise of AI and human collaboration has not only empowered individuals but has also created opportunities for collective intelligence. By pooling together the diverse knowledge and perspectives of humans, combined with AI's ability to process and analyze vast amounts of information, we can tackle complex problems with unprecedented efficacy.

However, this collaboration is not without its challenges. Ethical considerations, such as privacy, bias, and accountability, must be addressed to ensure that AI-integrated collaboration is fair, just, and inclusive. Looking forward, it is crucial to navigate this partnership ethically and responsibly to fully leverage the benefits of AI while mitigating potential risks.

As we delve deeper into the concept of AI-integrated human collaboration, we begin to witness the limitless possibilities that lie ahead. In the following chapters, we will explore the transformative potential of cooperation, the ethical implications of this partnership, and the way forward to reshape our future through collaboration.

So, let us continue to embark on this journey of "The AI-Integrated Human Evolution: From Competition to Cooperation," where we envision a world where humans and AI combine their strengths to shape a brighter and more harmonious future.

Chapter 6: The Philosophy of AI: Reflecting on the Impact and Ethics

As we delve deeper into the world of AI, it becomes crucial to reflect upon its profound impact on society, and the ethical questions that arise with its rapid advancement. In this chapter, we will explore the philosophy behind AI and challenge our understanding of intelligence, consciousness, and the boundaries of human existence.

One fundamental question that emerges is: what is intelligence? Traditionally, intelligence has been attributed to humans and other living beings. However, as AI continues to progress, machines are increasingly demonstrating capabilities that rival or surpass human intelligence in specific domains. This prompts us to question what it truly means to be intelligent and whether machines can exhibit a form of genuine intelligence.

Reflecting on consciousness also becomes crucial. Consciousness is often regarded as a defining characteristic of human beings, encompassing our subjective experiences, self-awareness, and the ability to perceive the world. Can machines possess consciousness? If so, what implications does this have for our understanding of personhood and the ethics surrounding AI?

Moreover, the philosophy of AI raises ethical considerations. As we develop AI systems capable of independent decision-making, we must address concerns such as bias, fairness, and accountability. The decisions an AI makes are influenced by the data it is trained on, which can inherently contain biases and prejudices. How can we ensure the ethical use of AI and guard against the perpetuation of discriminatory practices?

The potential impact of AI on employment and the economy is another important area of philosophical inquiry. As AI technology advances, the automation of various jobs becomes a reality. This raises questions about the distribution of wealth, the need for retraining and reskilling, and the role of work in our lives. How do we navigate the changing landscape of work in an AI-driven society while ensuring social and economic stability?

Reflecting on these philosophical questions helps us navigate the complexities of AI and make informed decisions about its development and deployment. It calls for interdisciplinary collaboration, involving not only computer scientists and engineers but also philosophers, ethicists, psychologists, and sociologists.

As AI continues to evolve, it is crucial to establish regulatory frameworks and ethical guidelines that address these philosophical questions. Open dialogue and public engagement become paramount in shaping the future of AI technology, ensuring that its benefits are maximized while minimizing the risks and potential harms.

Ultimately, the philosophy of AI invites us to examine our values, beliefs, and ethics in the context of a rapidly changing technological landscape. It challenges us to consider the potential consequences of our actions and to strive for responsible and beneficial AI development.

In the next chapter, we will explore the fascinating world of AI in popular culture, where science fiction has provided us with captivating narratives, creative possibilities, and cautionary tales. So let us continue our journey and dive into the realm of AI as portrayed in books, movies, and other forms of storytelling.

Chapter 7: AI in Popular Culture: Exploring Fantasies, Fears, and Reflections

Throughout history, humans have been captivated by stories that explore the possibilities and implications of artificial intelligence. In this chapter, we will embark on a journey through popular culture and delve into the ways AI has been portrayed, celebrated, and feared in books, movies, and other forms of storytelling.

Science fiction literature has long been a vehicle for exploring the philosophical and ethical implications of AI. From Mary Shelley's "Frankenstein" to Isaac Asimov's "I, Robot," authors have delved into themes of creation, power, and the definition of humanity. These stories challenge readers to ponder the nature of our relationship with intelligent machines and the potential consequences of our actions.

In the realm of cinema, AI has been both a fascinating subject and a source of anxiety. From iconic films like "Blade Runner" and "The Terminator" to contemporary works like "Her" and "Ex Machina," movies have depicted a wide range of AI scenarios. Some portray AI as potential saviors of humanity, offering solutions to our problems, while others depict them as existential threats, leading to the downfall of civilization.

Television series have also explored AI, often presenting complex and nuanced narratives. "Black Mirror" has gained popularity for its thought-provoking episodes that explore the dark side of technology, including AI-driven simulations, social credit systems, and automated decision-making. These stories serve as cautionary tales, reminding us of the ethical pitfalls and unintended consequences that AI can bring.

Beyond books and movies, AI has made its mark in video games, where virtual worlds offer unique opportunities for interaction with AI characters. Games like "Deus Ex" and "Detroit: Become Human" pose moral dilemmas, allowing players to navigate complex choices and determine the outcome of the story.

The portrayal of AI in popular culture serves not only as entertainment but also as a reflection of our collective hopes, fears, and aspirations. These narratives challenge us to contemplate the impact of AI on society, our understanding of humanity, and the ethical considerations that come with its advancement.

By critically engaging with these stories, we can gain insights into the potential risks and benefits of AI, enhancing our ability to shape its development in a responsible and thoughtful manner. Popular culture becomes a catalyst for public dialogue and engagement, encouraging us to explore the boundaries of AI and its integration into our everyday lives.

As we conclude this chapter, we must remember that AI in popular culture is but a reflection of our imagination and projection of our desires and concerns onto advancing technology. The real-world development of AI lies in the hands of scientists, engineers, policymakers, and society as a whole. It is up to us to shape the future of AI and ensure that it aligns with our values, respects our ethical principles, and enhances the well-being of humanity.

In the next chapter, we will dive into the practical applications of AI, exploring its impact across various industries and domains. So let us continue our journey and discover how AI is revolutionizing our world, one algorithm at a time.

AI in Action: Transforming Industries and Enhancing Human Potential

As the capabilities of artificial intelligence continue to advance, its applications across various industries and domains are becoming increasingly pronounced. The revolutionary potential of AI is transforming the way we work, communicate, and interact with the world around us. In this chapter, we will explore some of the key areas where AI is making a profound impact.

1. Healthcare: AI is revolutionizing healthcare by enabling early disease detection, personalized treatments, and improved patient care. Machine learning algorithms can analyze vast amounts of medical data, assist in diagnosis, and help predict patient outcomes. AI-powered robotics are also aiding in surgeries, providing precision and reducing the risk of human error.

2. Finance: In the financial sector, AI is instrumental in tasks such as fraud detection, risk assessment, and algorithmic trading. Autonomous trading systems leverage sophisticated algorithms to make rapid investment decisions, optimizing portfolio management and maximizing returns. AI-

powered chatbots and virtual assistants are also enhancing customer service and providing personalized financial advice.

3. Transportation: The transportation industry is undergoing a significant transformation with the help of AI. Autonomous vehicles are becoming a reality, promising safer and more efficient roadways. AI algorithms analyze traffic patterns, optimize route planning, and predict maintenance needs, contributing to reduced congestion and improved logistics.

4. Education: AI is reshaping the field of education by providing personalized learning experiences. Intelligent tutoring systems adapt to individual student needs and offer tailored instruction. Natural language processing enables automated language learning and translation, making education more accessible and inclusive.

5. Manufacturing: AI and robotics are redefining manufacturing processes, leading to higher efficiency and productivity. Smart factories leverage AI algorithms for predictive maintenance, quality control, and supply chain optimization. Collaborative robots, or cobots, work alongside humans, enhancing safety and enabling flexible production lines.

6. Agriculture: AI is empowering farmers with advanced analytics and precision farming techniques. Drones equipped with AI algorithms can collect data on crop health, soil conditions, and water management, enabling optimized resource allocation and higher yields. AI-powered systems aid in pest detection and control, reducing the need for harmful pesticides.

7. Customer Service: AI-powered chatbots and virtual assistants are transforming customer service experiences. Natural language processing allows these systems to understand and respond to customer queries, providing instant support and reducing wait times. AI algorithms analyze customer preferences and behavior, enabling personalized recommendations and targeted marketing campaigns.

8. Cybersecurity: As cyber threats continue to evolve, AI plays a vital role in combating them. AI algorithms can identify patterns, detect anomalies, and respond to potential security breaches. Autonomous security systems adapt to new threats in real-time, offering enhanced protection against cyber attacks.

The applications of AI are vast and continually expanding across numerous domains. From healthcare to finance, transportation to education, and manufacturing to customer service, AI is revolutionizing industries and transforming the way we live and work.

While there are significant benefits to be gained from AI, it is crucial to address ethical considerations surrounding privacy, bias, and equitable access. As AI continues to evolve, it becomes imperative to ensure transparency, accountability, and human oversight to mitigate potential risks and create responsible AI systems.

In the next chapter, we will discuss the future of AI and explore the possibilities and challenges that lie ahead. So, let's continue our journey into the exciting world of artificial intelligence.

Chapter 8: The Future of AI: Possibilities and Challenges

As we gaze into the future of artificial intelligence, we are met with a vast landscape of possibilities and challenges. The rapid advancements in AI technology are reshaping industries, empowering individuals, and revolutionizing the way we live our lives. Let's delve into some key areas where AI is expected to make profound strides in the coming years.

1. Natural Language Processing: AI systems will continue to evolve in their ability to understand and generate human language. We can expect more sophisticated voice assistants, chatbots, and language translation tools that are capable of engaging in natural, conversational interactions. This will facilitate seamless communication and enhance the overall user experience.

2. Autonomous Systems: The development of autonomous systems will be a significant focus in the years to come. We can expect to see further advancements in self-driving cars, drones, and robots that can perform complex tasks without human intervention. This will revolutionize industries such as transportation, delivery services, and logistics.

3. Advanced Healthcare Solutions: AI will continue to play a crucial role in the healthcare industry. We can anticipate the development of more accurate diagnostic tools, personalized treatment plans, and innovative solutions for disease prevention. AI algorithms will analyze vast amounts of medical data, leading to faster and more accurate diagnoses, and improved patient outcomes.

4. Enhanced Personalization: AI will enable highly personalized experiences across various domains, including entertainment, marketing,

and education. With the help of AI algorithms, businesses will be able to deliver tailored recommendations, customized advertisements, and individualized learning experiences. This will enhance customer satisfaction and improve engagement.

5. Robotics: Robotics will become increasingly sophisticated, capable of performing tasks previously reserved for humans. We can expect to see robots that can navigate complex environments, assist in household chores, and even provide companionship. Robotic advancements will transform industries and redefine the way we interact with machines.

While the future of AI holds great promise, it also comes with several challenges that need to be addressed.

1. Ethical Concerns: As AI becomes more pervasive, ethical considerations surrounding privacy, bias, and accountability become paramount. It is crucial to ensure that AI systems are designed and deployed in an ethically responsible manner, with transparency and fairness at their core.

2. Job Displacement: The adoption of AI and automation may lead to job displacement in certain industries. It is imperative to anticipate these changes and develop strategies to reskill and upskill workers to adapt to the evolving job market. Collaboration between humans and AI will be key to harnessing its full potential.

3. Data Privacy and Security: With AI relying heavily on data, there is a need for robust privacy and security measures to protect sensitive information. It is essential to establish strict regulations and standards to safeguard data and ensure that AI systems are not vulnerable to malicious attacks or misuse.

4. Human-AI Collaboration: Finding the right balance between human decision-making and AI assistance is crucial. It is important to establish frameworks for human oversight and ensure that AI systems are designed to augment human capabilities rather than replace them entirely.

As we navigate the path forward, it is crucial to address these challenges while embracing the transformative potential of AI. By fostering collaboration, prioritizing ethical considerations, and applying AI in a responsible manner, we can create a future where humans and AI coexist harmoniously, unlocking new possibilities and enhancing the human experience.

In the final chapter of this book, we will reflect on the journey we have taken through the world of AI, recap the key takeaways, and contemplate the future of this dynamic field. So, let's embark on our final chapter and conclude our exploration of artificial intelligence.

-

As we reach the final chapter of our exploration of artificial intelligence, let's take a moment to reflect on the journey we have taken and consider the future of this rapidly evolving field.

Throughout this book, we have delved into the various aspects of AI, from its history and fundamental concepts to its applications and potential impact on society. We have seen how AI has emerged as a powerful tool, transforming industries and enhancing our daily lives.

One of the key takeaways from our exploration is the importance of responsible and ethical AI development. As AI becomes increasingly integrated into our lives, it is essential that we prioritize transparency, fairness, and accountability. By designing AI systems that are explainable and unbiased, we can build trust and ensure that AI benefits all of humanity.

Another important aspect to consider is the role of human-AI collaboration. While AI has the potential to automate certain tasks and augment human capabilities, it is vital to maintain a balance. We must recognize that human judgment, creativity, and empathy are irreplaceable traits that can enhance AI systems and ensure that they align with our values and goals.

Looking ahead, there are several key considerations for the future of AI:

1. Continued Research and Development: AI is a field that is continually evolving, with new breakthroughs and innovations occurring regularly. Ongoing research and development are crucial to pushing the boundaries of AI and unlocking its full potential. Investment in AI education, interdisciplinary collaboration, and funding for research institutions will be essential to drive progress in the field.

2. Ethical Standards and Regulation: As AI becomes more prevalent in society, it is imperative to establish ethical standards and regulations to govern its development and use. This includes addressing privacy concerns, preventing algorithmic bias, and ensuring that AI systems are accountable

and transparent. Collaboration between policymakers, AI researchers, and industry leaders will be essential to develop comprehensive frameworks that promote ethical and responsible AI deployment.

3. Addressing Social Impact: AI has the potential to exacerbate existing societal issues or create new ones. It is important to consider the social impact of AI and anticipate any unintended consequences. This involves addressing issues such as job displacement, income inequality, and access to AI technology. By proactively addressing these challenges, we can mitigate potential negative effects and ensure that AI benefits all members of society.

4. Education and Upskilling: As AI continues to evolve, it is crucial to equip individuals with the skills and knowledge necessary to adapt to the changing job market. Providing education and upskilling opportunities will be vital to help people thrive in a world where AI plays an increasingly prominent role. This includes emphasizing critical thinking, problem-solving, and creativity, as these are skills that are uniquely human and complement the capabilities of AI.

In conclusion, our journey through the world of artificial intelligence has revealed its vast potential and the profound impact it can have on our lives. As we move forward, it is crucial to harness this potential responsibly, with a focus on ethics, collaboration, and addressing societal challenges.

By embracing the transformative power of AI while remaining mindful of its implications, we can shape a future where AI and humanity coexist harmoniously, advancing our collective progress and improving the well-being of individuals and societies as a whole.

Thank you for joining me on this exploration of artificial intelligence. I hope this book has provided you with valuable insights and sparked your curiosity about the many possibilities that AI holds. As AI continues to evolve, let us embark on this journey together, shaping a future where AI serves as a force for good.

Chapter 9: Exploring Transhumanism and Cutting-Edge Technologies

In this chapter, we will delve into the realm of transhumanism, a movement that seeks to enhance human capabilities through the integration of advanced technologies into our bodies and minds. We will explore some fascinating and potentially game-changing developments that are on the horizon, such as brain-computer interfaces and biohacking.

One of the most talked-about advancements in recent years is Neuralink, a company founded by entrepreneur Elon Musk. Neuralink aims to develop high-bandwidth brain-machine interfaces that can be implanted into the human brain, allowing for seamless communication between our minds and computer systems. This has the potential to revolutionize the way we interact with technology, opening up new possibilities in fields such as healthcare, communication, and entertainment.

Imagine a future where we can control our devices, access information, and even communicate with each other using nothing but our thoughts. Neuralink's technology could enable individuals with paralysis to regain control of their limbs or allow people with disabilities to overcome their limitations and fully participate in society. This fusion of technology and biology holds incredible promise for improving human lives in ways we cannot even fathom.

Another area that has gained attention is biohacking. Biohackers are individuals who experiment with their bodies, often using technology to enhance their physical or cognitive abilities. One example of biohacking is the implantation of devices, such as microchips, under the skin. These chips can be used for various purposes, such as replacing traditional car keys or credit cards.

Imagine a world where you no longer need to carry physical objects like keys or wallets. With a simple wave of your hand, doors open, payments are made, and access to various services is granted. This concept may seem like something out of a science fiction movie, but biohacking pioneers are already experimenting with these possibilities.

While these advancements hold tremendous potential, they also raise important ethical and privacy considerations. As we integrate technology deeper into our bodies, questions arise regarding data security, consent, and the potential for misuse. It is crucial that we have robust regulations and ethical frameworks in place to ensure that these technologies are developed and used responsibly, with the well-being and autonomy of individuals at the forefront.

As we look to the future, it is clear that the boundaries between humans and technology are becoming increasingly blurred. Transhumanism and the advancements it encompasses have the potential to redefine what it means to be human. It challenges us to imagine a world where our cognitive and physical abilities are enhanced beyond what was previously thought possible.

However, it is important to approach these advancements with a balanced perspective. While the integration of technology into our bodies offers exciting possibilities, it is essential to remember that our humanity lies not just in our capabilities, but in our values, emotions, and relationships.

As we navigate this uncharted territory, it is crucial to keep the principles of empathy, ethics, and inclusivity at the core of our decision-making process. By doing so, we can ensure that these transformative technologies enable us to enhance the human experience while preserving the qualities that make us uniquely human.

The future of transhumanism and these cutting-edge technologies is both exciting and uncertain. It will undoubtedly shape the course of human history and present us with new challenges and opportunities. As we cross the frontiers of what was once considered impossible, let us embark on this journey with a sense of wonder, responsibility, and an unyielding commitment to the betterment of humanity.

In this extended section, we will continue delving into transhumanism and the fascinating world of cutting-edge technologies that hold the potential to reshape the future as we know it. We will explore additional aspects of brain-computer interfaces, biohacking, and their profound implications on various facets of human existence.

A core focus within transhumanism is the development of brain-computer interfaces (BCIs) capable of seamlessly integrating our thoughts and mental processes with advanced computational systems. As the field progresses, we can envision a future where BCIs not only facilitate communication but also expand our cognitive abilities and enable direct access to vast networks of information.

Imagine being able to instantly retrieve knowledge from the internet, augment your mental capabilities, or even connect with the minds of others for collaborative problem-solving. These possibilities may sound straight out of a science fiction novel, but researchers and visionaries are actively working on making them a reality.

One of the exciting developments in this area is the concept of a "neural lace," a term coined by science fiction author Iain M. Banks. A neural lace refers to a network of microscopic sensors woven into the brain that can monitor and interact with neural activity. This technology could potentially enhance memory, cognition, and even facilitate direct communication between individuals.

Beyond BCIs, biohacking is another area that continues to captivate the imagination of numerous enthusiasts. Biohackers explore ways to augment their biology by conducting experiments on themselves or by actively participating in research and development projects. The motivation behind biohacking is often rooted in the pursuit of personal improvement, self-experimentation, and a desire to push the boundaries of human capabilities.

One fascinating aspect of biohacking is the endeavor to fuse technological devices with our bodies, allowing for unprecedented integration of hardware and biology. For instance, some individuals have experimented with sub-dermal implants that can monitor vital signs, track health data, or even serve as a form of sensory augmentation.

The possibilities of biohacking are vast, ranging from implants that enhance physical abilities to ones that enable new sensory experiences. For instance, imagine having an implant that allows you to sense electromagnetic fields or perceive ultraviolet light. Biohacking enthusiasts strive to explore these possibilities with a pioneering spirit, constantly pushing the boundaries of what is feasible.

With the advancements in transhumanism and cutting-edge technologies, there are also concerns and ethical considerations that must be addressed. Privacy and security, for instance, become critical issues when we integrate technology into our bodies and minds. Protecting the sensitive data generated by brain-computer interfaces or implantable devices becomes paramount in a future where our thoughts and physiological data can be directly accessed.

Furthermore, there are ethical dilemmas surrounding human enhancement. As we modify and augment our bodies and cognitive capabilities, questions of fairness, societal implications, and the potential for creating disparities arise. Striking a balance between individual freedom and the collective well-being of society poses significant challenges that must be carefully navigated as we embrace these emerging technologies.

It is essential to engage in ongoing ethical discussions, involving policymakers, scientists, philosophers, and the wider public, to ensure that the development and implementation of these technologies align with our shared values and priorities.

As we reflect on the potential trajectory of transhumanism and cutting-edge technologies, it becomes evident that humanity is standing at the threshold of a new era. These advancements hold immense promise, but also require responsible and thoughtful stewardship to ensure that they contribute to the betterment of our collective human experience.

The journey towards transhumanism challenges us to explore the interplay between technology, humanity, and our evolving understanding of what it means to be human. These advancements offer an unprecedented opportunity to improve the quality of life for individuals with disabilities, revolutionize healthcare, and enhance our cognitive abilities. However, it is crucial to acknowledge the potential risks, ethical considerations, and the need for transparent governance as we walk this uncharted path.

In embracing transhumanism and its related technologies, let us foster a culture of open dialogue, grounded in empathy and the recognition that our shared humanity remains at the core of all advancements and aspirations. By leveraging these technologies responsibly, ethically, and inclusively, we can build a future that is not only technologically advanced but also enhances the essence of being human.

Venturing into the realms of transhumanism and cutting-edge technologies, we open the door to a world of limitless possibilities and unimaginable opportunities. As we gaze into the future, the concept of futurism becomes an integral part of this exhilarating journey.

Futurism, at its core, is the study and anticipation of the future, fueled by an insatiable curiosity and an unyielding desire to explore the uncharted territories that lie ahead. It provides us with a framework to imagine, create, and shape a future that goes beyond the confines of our current realities, propelling us towards a new era of human existence.

The transhumanist movement, with its emphasis on human enhancement and the merging of biology with technology, aligns harmoniously with the principles of futurism. It challenges us to reimagine ourselves and our potential, igniting our imaginations and expanding the horizons of what it means to be human.

One fascinating aspect of futurism is the exploration of immortality and the quest to transcend the limitations of finite life. While achieving true immortality might still be beyond our grasp, the thirst for elongated lifespans and even the potential for rejuvenation is a tantalizing prospect. Advances in regenerative medicine, gene therapy, and nanotechnology offer glimpses of a future where we could extend our vitality and live healthier, more vibrant lives for decades longer.

In the realm of artificial intelligence (AI), futurism invites us to envision a future where machines possess human-like consciousness and surpass the boundaries of our own intelligence. The emergence of superintelligent AI could revolutionize not only our understanding of the universe but also our ability to solve complex problems, unravel the mysteries of existence, and pioneer innovations beyond our imagination.

Autonomous vehicles, an emblem of futurism, are set to redefine our transportation systems, presenting a vision of a future where road accidents become a relic of the past, and congestion is a distant memory. With the advent of flying cars, hyperloop systems, and advancements in sustainable transportation, we embark on a journey towards a utopian future where commuting is efficient, eco-friendly, and the concept of distance is redefined.

Space exploration represents yet another frontier that fuels the spirit of futurism. As we venture outwards into the cosmos, we become the pioneers of an era that could witness the colonization of other planets, the discovery of extraterrestrial life, and the unraveling of the mysteries that lie beyond our earthly domain.

Imagine a future where we establish colonies on Mars, harness the resources of asteroids, and engage in interstellar travel. The possibilities are astonishing, calling out to our insatiable curiosity and desire to explore the unexplored.

However, as we immerse ourselves in the excitement of futurism, it is essential to approach these advancements with an open mind, acknowledging the potential pitfalls, challenges, and ethical considerations that emerge along the way.

Addressing the social and economic implications of automation and AI, ensuring equitable access to these transformative technologies, and nurturing sustainability in the face of rapid advancements are crucial endeavors. The responsibility lies with us, as custodians of the future, to guide the course of progress, tempering ambition with wisdom, and fostering a future that is inclusive, just, and harmonious.

In this thrilling epoch of transhumanism and technological advancements, let us embrace futurism with unwavering optimism, curiosity, and the belief that humanity is destined for extraordinary achievements. Together, we can mold a future where the boundaries of possibility are continuously pushed to new frontiers, and the marvels of yesterday become the stepping stones to an era that surpasses even our wildest dreams.

Let us embark on this transformative journey, fueled by the spirit of exploration, ingenuity, and a profound belief in the incredible potential of human imagination. The future awaits, beckoning us to bring forth a world that is not only exciting but also embodies the very essence of what it means to be human.

Venturing into the realms of transhumanism and cutting-edge technologies, we open the door to a world of limitless possibilities and unimaginable opportunities. As we gaze into the future, the concept of futurism becomes an integral part of this exhilarating journey.

Futurism, at its core, is the study and anticipation of the future, fueled by an insatiable curiosity and an unyielding desire to explore the uncharted territories that lie ahead. It provides us with a framework to imagine, create, and shape a future that goes beyond the confines of our current realities, propelling us towards a new era of human existence.

The transhumanist movement, with its emphasis on human enhancement and the merging of biology with technology, aligns harmoniously with the principles of futurism. It challenges us to reimagine ourselves and our potential, igniting our imaginations and expanding the horizons of what it means to be human.

One fascinating aspect of futurism is the exploration of immortality and the quest to transcend the limitations of finite life. While achieving true immortality might still be beyond our grasp, the thirst for elongated lifespans and even the potential for rejuvenation is a tantalizing prospect. Advances in regenerative medicine, gene therapy, and nanotechnology offer glimpses of a future where we could extend our vitality and live healthier, more vibrant lives for decades longer.

In the realm of artificial intelligence (AI), futurism invites us to envision a future where machines possess human-like consciousness and surpass the boundaries of our own intelligence. The emergence of superintelligent AI could revolutionize not only our understanding of the universe but also our ability to solve complex problems, unravel the mysteries of existence, and pioneer innovations beyond our imagination.

Autonomous vehicles, an emblem of futurism, are set to redefine our transportation systems, presenting a vision of a future where road accidents become a relic of the past, and congestion is a distant memory. With the advent of flying cars, hyperloop systems, and advancements in sustainable transportation, we embark on a journey towards a utopian future where commuting is efficient, eco-friendly, and the concept of distance is redefined.

Space exploration represents yet another frontier that fuels the spirit of futurism. As we venture outwards into the cosmos, we become the pioneers of an era that could witness the colonization of other planets, the discovery of extraterrestrial life, and the unraveling of the mysteries that lie beyond our earthly domain.

Imagine a future where we establish colonies on Mars, harness the resources of asteroids, and engage in interstellar travel. The possibilities are astonishing, calling out to our insatiable curiosity and desire to explore the unexplored.

However, as we immerse ourselves in the excitement of futurism, it is essential to approach these advancements with an open mind, acknowledging the potential pitfalls, challenges, and ethical considerations that emerge along the way.

Addressing the social and economic implications of automation and AI, ensuring equitable access to these transformative technologies, and nurturing sustainability in the face of rapid advancements are crucial endeavors. The responsibility lies with us, as custodians of the future, to guide the course of progress, tempering ambition with wisdom, and fostering a future that is inclusive, just, and harmonious.

In this thrilling epoch of transhumanism and technological advancements, let us embrace futurism with unwavering optimism, curiosity, and the belief that humanity is destined for extraordinary achievements. Together, we can mold a future where the boundaries of possibility are continuously pushed to new frontiers, and the marvels of yesterday become the stepping stones to an era that surpasses even our wildest dreams.

Let us embark on this transformative journey, fueled by the spirit of exploration, ingenuity, and a profound belief in the incredible potential of human imagination. The future awaits, beckoning us to bring forth a world that is not only exciting but also embodies the very essence of what it means to be human.

Chapter 10: Ethical Considerations and learning in a Technological Future

As we dive deeper into the realm of futurism, it is crucial to pause and reflect upon the ethical considerations that arise from the rapid advancement of technology. The future we envision must not only be exciting, but also morally sound, respectful of individual rights, and cognizant of the potential risks that lie ahead.

In this chapter, we explore the ethical landscapes that will shape our technological future and examine the dilemmas that demand our attention and thoughtful contemplation.

One pressing issue is the ethical use of artificial intelligence. As AI becomes increasingly powerful and autonomous, questions arise regarding the accountability and transparency of its decision-making processes. We must grapple with the conundrum of ensuring that AI operates within ethical boundaries, respects human values, and avoids bias or discriminatory practices.

Moreover, the impact of AI on employment also necessitates ethical considerations. As automation continues to disrupt various industries, we must address the potential economic and social consequences of job displacement. Fostering inclusive growth and implementing measures to support individuals affected by automation will be paramount in creating a future that is fair and equitable.

Another ethical domain that warrants our attention is privacy and data protection. In this digital age, we generate vast amounts of personal data, which when in the wrong hands, can infringe upon our privacy and even be used against us. Striking a balance between technological progress and safeguarding individual privacy becomes imperative to protect fundamental human rights.

Additionally, the advent of biotechnology and genetic engineering presents us with new ethical predicaments. As we gain the ability to manipulate and enhance the human genome, questions of consent, equity, and the boundaries of what is considered "natural" come to the forefront. Responsible governance and an ethical framework must guide our decisions to ensure the ethical use of these powerful technologies.

Furthermore, the potential consequences of climate change and ecological destruction necessitate ethical action. As we push the limits of innovation, developing sustainable technologies and ensuring the responsible use of resources must be central tenets of our journey towards the future. It is our duty to safeguard the planet for generations to come and mitigate the harm caused by human activities.

The future demands that we approach these ethical challenges with an open mind, engaging in collaborative dialogues that incorporate diverse perspectives. It calls for global cooperation, encompassing policymakers, scientists, ethicists, and individuals from all walks of life.

As we navigate the ever-expanding horizon of technology, let us not forget our moral compass. By intertwining our passion for progress with steadfast ethical principles, we can forge a path towards a future that is not only exciting and transformative but also preserves the cherished values that define our shared humanity.

In the next chapter, let us delve into the significance of creativity and imagination in shaping our future. Join me as we explore the boundless realms of human ingenuity and uncover the ways in which our creative spirit will propel us towards a thriving and harmonious tomorrow.

Lifelong learning plays a crucial role in a rapidly changing world. As the pace of technological advancements accelerates and industries evolve, individuals must continuously acquire new knowledge and skills to remain competitive and adaptable. The importance of traditional education is undeniable, but it can no longer be seen as a one-time event that ends with graduation.

In a changing world, new challenges and opportunities arise constantly. Lifelong learning enables us to stay relevant and keep up with the demands of our professional and personal lives. It helps us embrace new technologies, approaches, and ideas, enabling us to enhance our problem-solving abilities and become more innovative.

One key aspect of lifelong learning is its ability to foster a growth mindset. By actively seeking new knowledge and skills, we cultivate an attitude of curiosity, perseverance, and flexibility. This mindset enables us to embrace change and uncertainty, rather than fear or resist it. It empowers us to view challenges as opportunities for growth and to continuously improve ourselves by acquiring new perspectives and insights.

Furthermore, lifelong learning expands our horizons and broadens our understanding of the world. It encourages us to engage with diverse ideas, cultures, and perspectives, promoting empathy and understanding. In a globalized world, where interconnectedness is the norm, being culturally aware and sensitive is essential for effective communication and collaboration.

Additionally, lifelong learning promotes personal development and well-being. It provides opportunities for self-reflection, self-improvement, and personal growth. As we learn new things and acquire new skills, we build confidence in ourselves and our abilities. This, in turn, boosts our overall well-being and enhances our sense of fulfillment and purpose in life.

In summary, lifelong learning is crucial in a changing world because it enables us to adapt, innovate, and thrive. It empowers individuals with the knowledge, skills, and mindset necessary to navigate the complex challenges of our time. By embracing lifelong learning, we can confidently face the future and actively shape it in ways that lead to personal and collective success.

Lifelong learning is an ever-evolving concept that embraces the idea of continuous personal and professional growth. As the world becomes increasingly interconnected and new technologies emerge, the potential for humans to access machine learning holds promising possibilities.

Machine learning, a subset of artificial intelligence, has seen rapid advancements in recent years. It involves the development of algorithms that enable computers to learn and make predictions or decisions without being explicitly programmed. While machines currently dominate in certain areas, such as data analysis and pattern recognition, humans have the advantage of creativity, critical thinking, and emotional intelligence.

The intersection of human learning and machine learning opens up exciting new opportunities. Imagine a future where humans can access machine learning algorithms to accelerate their language learning abilities. Instead of spending years acquiring fluency in a new language, individuals could potentially tap into a vast database of language patterns and nuances, expediting the language-learning process.

This fusion of human and machine learning could also enhance problem-solving capabilities. By combining human ingenuity with the computational power of machine learning, individuals would be better equipped to tackle complex challenges and innovate in various domains. The synergy of human intuition and machine-powered data analysis could lead to ground-breaking discoveries and solutions.

However, it is essential to consider the ethical implications and potential pitfalls of this convergence. Privacy, security, and fairness become critical concerns. Ensuring that access to machine learning technologies is available to all and that its development aligns with ethical principles will be vital in this future landscape.

In this context, lifelong learning takes on a new dimension. It becomes not only about continuous personal growth but also about navigating the dynamic relationship between humans and technology. Individuals must cultivate a balance between leveraging machine learning capabilities and preserving their unique human qualities.

Moreover, as new technologies emerge, the need for learning and adaptability becomes greater. Lifelong learners will need to acquire not only new technical skills but also the ability to understand and work alongside intelligent machines. This requires a mindset of curiosity, agility, and adaptability—a mindset that embraces the ever-changing nature of our world and actively seeks to bridge the gap between human and machine.

In conclusion, lifelong learning holds tremendous value as we consider the possibilities of accessing machine learning in the future. By embracing the fusion of human and machine capabilities, individuals can potentially accelerate their learning processes and enhance their problem-solving abilities. However, it is crucial to approach this convergence with ethical considerations in mind and strive for a balanced integration that preserves the unique qualities and potential of human beings.

Chapter 11: Transcending Individualism

A. The pitfalls of hyper-competition and individualistic mindsets

In a world that has long celebrated individualism and competition, it is important to acknowledge the potential pitfalls of such mindsets. While competition can drive innovation and progress, an exclusive focus on personal success and gains can lead to a fragmented society and systemic inequalities. Hyper-competition can erode collaboration, hinder collective problem-solving, and isolate individuals from one another.

When individuals are solely focused on their own goals, cooperation and empathy often take a back seat. This mentality can impede progress in areas that require collective effort, such as sustainable development, healthcare, and climate change mitigation. Moreover, the pressures of hyper-competition can lead to stress, burnout, and a diminished sense of well-being.

B. The philosophy of cooperation and mutual benefit

Recognizing the limitations of individualism, societies have long embraced the philosophy of cooperation and mutual benefit. When individuals come together, share resources, and collaborate, collective achievements can be far greater than what can be accomplished alone. Cooperation encourages problem-solving, fosters innovation, and promotes a sense of community and shared purpose.

Cooperation can take various forms, from local community initiatives to international collaborations. By valuing cooperation, societies can shift their focus from an individual's success to the well-being and progress of the collective. This shift encourages inclusive growth and ensures that the benefits of progress are shared more equitably.

C. Case studies of successful collaborations and collective achievements

Throughout history, there have been numerous examples of civilizations that have exemplified the power of cooperation and achieved greatness through collective efforts. One prominent example is the construction of the pyramids in ancient Egypt. These monumental structures required the collaboration of thousands of workers, engineers, and architects over extended periods of time. It was through their collective expertise, coordination, and shared vision that the pyramids were built, serving as lasting testaments to the greatness of ancient Egyptian civilization.

Another example is the International Space Station (ISS), a remarkable collaboration between multiple nations. The ISS is a symbol of global cooperation, as countries set aside political differences and worked together to create a habitable laboratory in space. This extraordinary achievement serves as a reminder of the potential for collective efforts to transcend boundaries and achieve monumental goals.

Furthermore, in the field of scientific research, collaboration has been crucial in advancing knowledge and making groundbreaking discoveries. Large-scale collaborations like the Human Genome Project, which involved researchers from around the world, led to the mapping of the entire human genome. This collaborative effort has had far-reaching implications for medical research and personalized healthcare.

As societies evolve, the possibility of transcending individualism and embracing cooperation at a new level arises. While speculating about the future, concepts such as human-machine integration, often referred to as "cyborgs," or the emergence of highly advanced artificial intelligence, labeled "AI-sapiens," have been explored. In these scenarios, the cooperation between humans and machines becomes an essential element in achieving new levels of progress and understanding.

In conclusion, transcending individualism and promoting cooperation is essential for the continued progress and well-being of society. By recognizing and mitigating the pitfalls of hyper-competition, societies can foster collaboration, empathy, and a shared sense of purpose. Drawing inspiration from historical and contemporary examples of successful collaborations, we can envision a future where collective achievements propel humanity forward, whether through ambitious construction projects like the pyramids, global scientific endeavors like the Human Genome Project, or even transformative partnerships with advanced technologies in the form of "cyborgs" or "AI-sapiens."

In the quest to understand the potential risks faced by a species, it becomes evident that lack of cooperation among various species has led to their extinction in the past. Cooperation and collaborative efforts are essential for survival and long-term well-being.

As humans, we face a multitude of existential risks including nuclear war, climate disasters, and other global challenges. These risks emphasize the critical importance of cooperation and collective action in addressing and mitigating such threats. By working together, nations can develop strategies and frameworks to prevent conflicts and reduce the likelihood of catastrophic events. Similarly, global cooperation is crucial in tackling climate change, where the actions of individual countries alone may not be sufficient.

The emergence of artificial intelligence (AI) complicates our understanding of existential risks. While AI has the potential to bring about significant advancements in various fields, it also carries certain risks. Concerns have been raised regarding AI surpassing human capabilities and becoming uncontrollable or even hostile. However, it is important to note that the current level of AI development does not pose an immediate existential threat.

Debates regarding the risks of AI development are ongoing. Some argue that AI can be harnessed to address existing existential risks more effectively. For example, AI can assist in climate modeling, resource management, and disaster response. It can also contribute to advances in medicine, robotics, and other fields that improve the well-being of humanity.

Nevertheless, it is crucial to exercise caution in AI development. Responsible and ethical deployment of AI technologies is vital, incorporating sufficient safeguards and regulations to prevent misuse or unintended consequences. Initiatives are being undertaken to address these concerns, including the development of ethical frameworks and governance models for AI.

Regarding the challenge of reversing AI development, it is true that AI has become deeply integrated into our modern society. Many industries and businesses rely on AI technologies for various purposes, such as automation, data analysis, and customer service. However, as we better understand the risks and consequences of AI, there is a growing emphasis on developing mechanisms for AI safety and regulation.

The responsible development and deployment of AI necessitates collaboration among various stakeholders, including policymakers, researchers, and industry leaders. A balanced approach is required, maximizing the benefits of AI while mitigating potential risks. This can involve establishing ethical guidelines, implementing robust technical standards, and fostering open dialogue and cooperation across society.

In conclusion, the exploration of transcending individualism and promoting cooperation addresses the vital importance of collaboration in avoiding existential risks and achieving collective progress. While both lack of cooperation and the risks associated with AI development can be threatening, it is through cooperation, responsible governance, and collective problem-solving that we can navigate these challenges and create a future that benefits all of humanity.

Cooperation is not only essential for addressing existential risks, but it also holds immense potential in various aspects of human society. Let's explore how cooperation can benefit the economy, global health, and the overall happiness of humanity.

In terms of the economy, cooperation plays a significant role in promoting growth, stability, and prosperity. Collaborative efforts between businesses, industries, and even nations can lead to the exchange of resources, expertise, and technology. By pooling resources and sharing knowledge, economies can become more efficient and innovative. Cooperation also fosters healthy competition, encouraging businesses to strive for excellence and provide better products or services. Additionally, collaborations between different sectors can lead to the creation of new industries and job opportunities, contributing to economic development.

Cooperation is also instrumental in improving global health outcomes. In an interconnected world, infectious diseases can spread rapidly across borders. International cooperation is crucial in detecting, preventing, and containing such health crises. Collaborative efforts, such as sharing data, research, and resources, enable countries to respond effectively to outbreaks and pandemics. Cooperation can lead to the development of vaccines, treatments, and effective healthcare systems that benefit all nations. Furthermore, global cooperation in healthcare can address larger challenges, including access to affordable medications, equitable distribution of healthcare resources, and tackling major health issues such as non-communicable diseases.

When it comes to happiness for humanity as a whole, cooperation plays a vital role in fostering a sense of community, belonging, and social support. By working together, individuals and communities can address common challenges and achieve common goals. Cooperation promotes trust, empathy, and a sense of shared responsibility, all of which contribute to societal well-being and overall happiness. It can lead to the formation of strong social bonds, increased social capital, and a sense of collective identity. The collaborative efforts to support and care for each other create a positive environment that enhances the happiness and quality of life for individuals and communities.

While competition serves a purpose in nature and can drive progress, it is indeed possible to reach a higher level by combining it with cooperation. Cooperation does not mean the absence of competition but rather supplementing it with collaboration and mutual benefit. Instead of focusing solely on individual success, cooperation encourages collective advancement. It allows individuals, businesses, and nations to harness their unique strengths and capabilities while collaborating on areas of shared interest. By combining competition with cooperation, we can create an environment where individuals thrive, businesses innovate, and societies progress harmoniously.

In summary, cooperation brings forth numerous benefits to various aspects of human society. It fuels economic growth, enables efficient responses to global health challenges, and contributes to the overall happiness and well-being of humanity. By combining competition with cooperation, we can maximize the potential for progress and create a future built on collaboration, shared success, and collective flourishing.

In addition to the benefits mentioned earlier, cooperation can have several other positive impacts on humanity as a whole.

1. Sustainable Development: Cooperation is crucial for achieving sustainable development goals. As we face challenges like climate change, resource depletion, and environmental degradation, collaboration among governments, organizations, and communities becomes essential. Through cooperation, we can collectively work towards implementing environmentally-friendly practices, promoting renewable energy sources, and conserving natural resources for future generations.

2. Peace and Conflict Resolution: Cooperation plays a vital role in fostering peace and resolving conflicts. By promoting dialogue, understanding, and compromise, cooperation can help build bridges between nations and communities. Diplomatic efforts, negotiations, and international treaties are successful examples of how cooperation can lead to peaceful resolutions and prevent wars or conflicts.

3. Innovation and Knowledge Sharing: Cooperation accelerates innovation and the exchange of knowledge. When individuals and organizations come together, they can pool their expertise, resources, and ideas, leading to breakthrough innovations and scientific advancements. Collaborative research, academic partnerships, and open-source initiatives are examples of how cooperation can drive progress and benefit society as a whole.

4. Social Equality and Inclusion: Cooperation can contribute to creating a more equitable and inclusive society. By working together and embracing diversity, we can address social inequalities, combat discrimination, and promote equal opportunities for all. Cooperative efforts can focus on

empowering marginalized communities, promoting education, and reducing economic disparities, fostering a more just and inclusive society.

5. Resilience and Crisis Management: Cooperation is essential during times of crisis and emergencies. Whether it's natural disasters, economic downturns, or public health crises, collaborative efforts enable countries and communities to respond quickly and effectively. By sharing resources, expertise, and support, cooperation strengthens our ability to mitigate the impact of crises and recover more efficiently.

In summary, cooperation has a profound impact on several aspects of human society. It facilitates sustainable development, peacebuilding, innovation, social equality, and resilience. By recognizing the power of cooperation and fostering collaborative relationships, we can address global challenges and create a brighter and more prosperous future for humanity.

Here are a few more points on the benefits of cooperation:

1. Education and Knowledge Transfer: Cooperation in the field of education allows for the exchange of knowledge, best practices, and educational resources. Collaborative efforts between educational institutions and organizations can facilitate the transfer of expertise and promote lifelong learning. Cooperation also enables access to educational opportunities for individuals who may not otherwise have access, promoting inclusivity and equal opportunity in education.

2. Resource Allocation and Efficiency: Cooperation helps maximize the efficient allocation of resources. By sharing costs, infrastructure, and expertise, organizations and communities can reduce redundancy and optimize resource utilization. This can translate into cost savings, increased productivity, and improved efficiency.

3. Cultural Exchange and Understanding: Cooperation fosters cultural exchange and understanding among diverse communities. Through collaborations, individuals from different backgrounds can come together, share their experiences, and learn from one another. This promotes tolerance, empathy, and appreciation for different cultures, leading to a more inclusive and harmonious society.

4. International Relations and Diplomacy: Cooperation is essential for building strong international relations and maintaining peace. Diplomatic collaborations and international alliances create platforms for dialogue,

negotiation, and conflict resolution. These cooperative efforts help prevent misunderstandings, minimize tensions, and promote mutual understanding and respect between nations.

5. **Personal Growth and Fulfillment:** Cooperation provides individuals with opportunities for personal growth and fulfillment. Working together towards a common goal fosters a sense of purpose, camaraderie, and accomplishment. It also promotes teamwork, leadership skills, and the development of social connections, leading to personal satisfaction and a sense of belonging.

6. **Environmental Conservation:** Cooperation is crucial in addressing environmental challenges and conserving natural resources. Collaborative efforts between governments, organizations, and individuals can lead to the preservation of ecosystems, the protection of endangered species, and the promotion of sustainable practices.

In summary, cooperation has wide-ranging benefits including education, resource efficiency, cultural understanding, international relations, personal growth, and environmental conservation. By embracing cooperation, we can create a more interconnected and thriving world that values collaboration and shared prosperity.

7. **Economic Growth and Prosperity:** Cooperation is one of the key drivers of economic growth and prosperity. When individuals, businesses, and countries collaborate, they can combine their resources, skills, and expertise to create more innovative products and services. This leads to increased productivity, job creation, and overall economic development.

8. **Humanitarian Aid and Disaster Relief:** Cooperation is crucial in providing humanitarian aid and disaster relief to those in need. When countries and organizations work together, they can pool their resources, expertise, and manpower to respond effectively to emergencies, provide assistance to affected communities, and facilitate long-term recovery efforts.

9. **Social Support and Community Building:** Cooperation plays a vital role in building strong and supportive communities. By working together, individuals can provide emotional support, share resources, and contribute to the well-being of others. Whether it's through volunteering, community organizations, or mutual aid groups, cooperation fosters a sense of togetherness and creates a safety net for those in need.

10. Conflict Prevention and Resolution: Cooperation is instrumental in preventing conflicts and resolving existing ones. Through collaboration, countries and communities can identify common interests, address grievances, and find mutually beneficial solutions. This proactive approach to conflict prevention can help mitigate tensions, promote stability, and avoid the devastating consequences of conflicts.

11. Improved Decision-Making: Cooperation allows for diverse perspectives to be considered when making important decisions. By involving multiple stakeholders and fostering inclusive decision-making processes, cooperation leads to better outcomes that take into account a wider range of interests and concerns.

12. Innovation and Entrepreneurship: Cooperation fosters innovation and entrepreneurship by creating an ecosystem that supports the exchange of ideas and collaboration between individuals and organizations. Collaborative networks, incubators, and innovation hubs provide opportunities for people to connect, share knowledge, and work together to develop new technologies, products, and services.

13. Health and Scientific Advancements: Cooperation is vital in the field of healthcare and scientific research. By collaborating on research projects, sharing data, and coordinating efforts, scientists and healthcare professionals can make breakthrough discoveries, develop new treatments and vaccines, and improve public health outcomes.

In summary, cooperation has numerous benefits, including economic growth, humanitarian aid, community building, conflict resolution, improved decision-making, innovation, and advancements in healthcare and science. By fostering a spirit of cooperation, we can create a more inclusive, resilient, and prosperous world for everyone.

Cooperation plays a significant role in driving innovation and progress. Here are a few ways in which cooperation facilitates these advancements:

1. Sharing of Resources and Expertise: Cooperation allows individuals and organizations to combine their resources, skills, and expertise. By working together and pooling their knowledge and resources, they can accomplish far more than they would individually. This collaboration encourages the exchange of ideas, best practices, and diverse perspectives, leading to innovative solutions and progress.

2. Cross-Disciplinary Collaboration: Cooperation enables individuals from different fields and disciplines to come together and collaborate. This cross-pollination of ideas and expertise often leads to breakthrough innovations. When people with varying backgrounds and areas of specialization collaborate, they can approach problems from different angles and leverage their collective knowledge to find novel solutions.

3. Open Innovation: Cooperation promotes open innovation, where organizations and individuals actively seek external input and collaborate with external partners. By engaging with customers, suppliers, research institutions, and other stakeholders, organizations can tap into a broader range of ideas and perspectives. This openness encourages the flow of information, fosters creativity, and accelerates the pace of innovation.

4. Collaborative Research and Development: Cooperation in research and development (R&D) activities allows for the sharing of scientific knowledge, data, and infrastructure. This collaborative approach to R&D enables researchers and scientists to build upon each other's work, validate findings, and make faster progress. It also helps reduce duplication of efforts and costs associated with individual R&D projects.

5. Supportive Ecosystems: Cooperation is nurtured in supportive ecosystems that foster innovation. Innovation hubs, research institutions, and entrepreneurial communities provide platforms where individuals and organizations can connect, collaborate, and share ideas. These ecosystems offer a supportive environment that encourages risk-taking, experimentation, and the exchange of knowledge, ultimately leading to innovation and progress.

6. Access to Funding and Resources: Cooperation often facilitates access to funding and resources necessary for innovation. By collaborating on projects or forming partnerships, individuals and organizations can leverage shared resources and attract investment. This access to funding and resources brings together the necessary ingredients for innovation, such as research infrastructure, talent, and capital.

7. Market Expansion and Commercialization: Cooperation among businesses can lead to the expansion of markets, increased customer reach, and accelerated commercialization of new products and services. Through partnerships and collaborations, companies can combine their strengths, access new markets, and leverage each other's distribution networks. This collaborative approach accelerates the time to market and enhances the likelihood of success.

In summary, cooperation fuels innovation and progress by promoting the sharing of resources and expertise, encouraging cross-disciplinary collaboration, enabling open innovation, facilitating collaborative research and development, nurturing supportive ecosystems, providing access to funding and resources, and facilitating market expansion and commercialization. These cooperative efforts yield groundbreaking innovations, drive economic growth, and contribute to societal progress.

Countries can collaborate in various ways to address climate change:

1. International Climate Agreements: Countries can come together to negotiate and sign international agreements aimed at curbing greenhouse gas emissions and mitigating climate change. The most prominent example is the Paris Agreement, where countries pledged to limit global warming to well below 2 degrees Celsius above pre-industrial levels. By participating in such agreements, countries commit to taking action and reporting progress on emissions reductions.

2. Information and Data Sharing: Collaboration involves sharing scientific research, data, and best practices related to climate change. Countries can share information on successful policies, technological innovations, and strategies for adaptation and resilience. By sharing knowledge, countries can learn from each other's experiences and avoid reinventing the wheel.

3. Financial Support: Developed countries can provide financial support to developing nations to help them transition to low-carbon economies and adapt to the impacts of climate change. This assistance can come in the form of grants, loans, and investments in renewable energy, sustainable infrastructure, and climate resilience projects. Financial cooperation plays a crucial role in ensuring that all countries have the necessary resources to address climate change effectively.

4. Technology Transfer: Developed countries can facilitate the transfer of clean and sustainable technologies to developing nations. This can be done through partnerships, knowledge exchange, and capacity building initiatives. By sharing technological advancements, countries can accelerate the adoption of low-carbon solutions and help bridge the technology gap between developed and developing nations.

5. Collaborative Research and Innovation: Countries can collaborate on research and development projects focused on climate mitigation and adaptation. Joint funding initiatives, scientific partnerships, and

knowledge-sharing platforms can facilitate collaborative efforts to develop and deploy sustainable technologies, promote renewable energy, and find innovative solutions to climate challenges.

6. Carbon Markets and Trading: Countries can participate in carbon markets or establish emissions trading systems, creating a mechanism for the exchange of emission allowances and the incentivization of emissions reductions. These market-based approaches can help countries meet their emission reduction targets more cost-effectively and encourage investment in clean technologies.

7. Awareness and Education: Collaboration in the form of educational campaigns, public awareness initiatives, and knowledge-sharing platforms can raise awareness about climate change and foster a sense of collective responsibility. By informing and engaging citizens, countries can create a supportive environment for climate action and encourage individuals to adopt sustainable practices.

Effective collaboration on a global scale is crucial for addressing climate change comprehensively. By combining efforts, sharing information, providing financial and technological support, and fostering innovation, countries can make significant progress in mitigating and adapting to the impacts of climate change.

Chapter 12: The Power of Mind-Body Integration

In this chapter, we delve into the fascinating realm of mind-body integration and explore the incredible advancements in neurotechnology and neural interfaces that have revolutionized our understanding of the human brain and its connection to the body. We also delve into the profound implications of enhancing cognitive capabilities and expanding consciousness, as well as the importance of cultivating well-being and balancing physical and mental well-being.

A. Exploring advancements in neurotechnology and neural interfaces:

We begin by examining the cutting-edge field of neurotechnology, which encompasses a wide range of technologies designed to interface with the human brain. From brain-computer interfaces (BCIs) to neuroprosthetics, these advancements have the potential to revolutionize healthcare, rehabilitation, and human performance. We discuss the development of non-invasive and invasive techniques, exploring their applications in restoring mobility to paralyzed individuals, aiding in stroke rehabilitation, and even augmenting cognitive abilities.

We delve into the remarkable progress made in understanding how the brain functions through techniques such as electroencephalography (EEG), functional magnetic resonance imaging (fMRI), and transcranial magnetic stimulation (TMS). These techniques allow us to study neural activity and connectivity, providing unprecedented insights into the workings of the mind.

B. Enhancing cognitive capabilities and expanding consciousness:

Next, we explore the concept of enhancing cognitive capabilities and expanding consciousness. We examine the potential of neurotechnological interventions, such as neurofeedback, cognitive training, and brain stimulation, to improve memory, attention, and learning abilities. We discuss the ethical implications of cognitive enhancement and the importance of responsible use of these technologies.

Furthermore, we explore the frontiers of consciousness exploration, discussing practices such as meditation, mindfulness, and psychedelic-assisted therapy. We delve into the scientific research that has highlighted the transformative effects of these practices on mental well-being, creativity, and spiritual experiences. We also address the challenges and controversies surrounding the integration of altered states of consciousness into mainstream medical and psychological therapies.

C. Cultivating well-being and balancing physical and mental well-being:

Finally, we emphasize the significance of cultivating well-being and achieving a harmonious balance between physical and mental health. We explore the interplay between the mind and the body, discussing the impact of mental states, emotions, and psychological well-being on physical health and vice versa.

We delve into various approaches to promoting well-being, including exercise, nutrition, stress reduction techniques, and the power of positive psychology. We also discuss the growing body of evidence supporting the mind-body connection and highlight the importance of holistic approaches to healing and self-care.

Throughout this chapter, we uncover the exciting possibilities that arise from the integration of mind and body. By embracing advancements in neurotechnology, expanding our consciousness, and fostering well-being, we have the potential to unlock new dimensions of human potential and lead healthier, more fulfilling lives.

In optimizing brain function, the role of nutrition cannot be overstated. The brain is an incredibly complex organ that requires a steady supply of nutrients to function efficiently. In this section, we will delve into the importance of nutrition and how it impacts various aspects of brain health and cognitive function.

1. Macronutrients: The brain relies heavily on a consistent supply of glucose, obtained from carbohydrates, for energy. Consuming complex carbohydrates, such as whole grains, fruits, and vegetables, provides a slow and steady release of glucose, promoting sustained mental energy throughout the day. Additionally, adequate intake of healthy fats, found in foods like avocados, nuts, and fatty fish, is crucial for maintaining the integrity of brain cell membranes and supporting cognitive function.

2. Micronutrients: Micronutrients, including vitamins and minerals, play a vital role in brain health. Some essential nutrients for brain function include:

- Omega-3 fatty acids: Found in fatty fish, walnuts, and flaxseeds, these healthy fats have been linked to improved cognitive function and reduced risk of age-related cognitive decline.

- B vitamins: B vitamins, such as B6, B12, and folate, are involved in the synthesis of neurotransmitters that regulate mood and cognitive processes. Good sources include leafy greens, legumes, and lean meats.

- Antioxidants: Antioxidants, found in colorful fruits and vegetables, help protect the brain from oxidative stress and inflammation, which can contribute to cognitive decline and neurodegenerative diseases.

- Minerals: Key minerals like iron, zinc, and magnesium are essential for proper brain function. Iron, found in meat, beans, and leafy greens, is necessary for oxygen transport to the brain. Zinc, present in oysters, beef, and pumpkin seeds, is involved in memory and learning processes. Magnesium, found in nuts, seeds, and whole grains, plays a crucial role in synaptic plasticity and cognitive function.

3. Hydration: Proper hydration is critical for optimal brain function. Even mild dehydration can impair cognitive abilities, attention, and memory. It

is important to drink enough water throughout the day to stay adequately hydrated.

4. Gut-Brain Connection: Emerging research highlights the influence of the gut microbiome on brain health. A healthy gut microbiome, which can be nourished through a balanced and varied diet rich in fiber, prebiotics, and fermented foods, supports brain function and may help mitigate the risk of certain mental health conditions.

5. Specific Diets: Various dietary approaches, such as the Mediterranean diet, DASH diet, and MIND diet, have been associated with better cognitive health and reduced risk of cognitive decline. These diets emphasize whole, nutrient-dense foods, including fruits, vegetables, whole grains, lean proteins, and healthy fats.

It's important to remember that nutrition is just one aspect of maintaining brain health. A holistic approach that incorporates regular exercise, adequate sleep, stress management, and cognitive stimulation is essential for optimizing brain function. Consulting with a healthcare professional or Registered Dietitian can provide personalized guidance based on individual needs and goals.

While there are various substances known as nootropics that have shown promising potential in enhancing cognitive performance, it's important to note that the field of cognitive enhancement is still evolving, and many of these substances require further research to fully understand their efficacy and safety. Here, I will discuss a few popular substances:

1. Dihexa: Dihexa is a peptide that has shown promise in animal studies for its neuroprotective and cognitive-enhancing effects. It is believed to improve synaptic connectivity and promote the growth of new neurons. However, it is important to emphasize that the research on dihexa is still in its early stages, and its effects on humans are not yet well-understood.

2. Racetams: Racetams are a class of compounds, including piracetam and aniracetam, that have been studied for their potential cognitive-enhancing effects. These compounds are thought to modulate neurotransmitter activity and promote better cognitive function. However, while some individuals report positive effects, the scientific evidence for their significant benefits in healthy individuals remains limited.

3. Modafinil: Modafinil is a wakefulness-promoting agent commonly used to treat sleep disorders like narcolepsy. It has gained attention as a

potential cognitive enhancer due to its ability to improve alertness and concentration. While modafinil is known to have effects on cognition, it is a prescription medication with potential side effects, and its use should be supervised by a healthcare professional.

As for future mind enhancers or the possibility of "limitless pills," it is challenging to predict with certainty what breakthroughs may occur. However, ongoing research is exploring various avenues, including the development of new substances, advanced brain-machine interfaces, and innovative therapies. Scientists are researching novel compounds, such as ampakines, BDNF modulators, and selective serotonin reuptake inhibitors (SSRIs), to enhance neuroplasticity, memory, and cognitive function. Additionally, advancements in non-invasive brain stimulation techniques, such as transcranial magnetic stimulation (TMS) and transcranial direct current stimulation (tDCS), show promise in modulating brain activity and potentially enhancing cognition.

Nevertheless, it's important to approach cognitive enhancement with caution and skepticism. Ethical considerations, potential risks, long-term effects, and individual variances in response should be taken into account. It is always advisable to consult with a healthcare professional before considering any nootropic or cognitive-enhancing substance.

While the idea of a "limitless pill" capturing the capabilities depicted in fictional works is captivating, it's crucial to maintain realistic expectations and prioritize holistic approaches to cognitive wellness, such as a balanced diet, regular exercise, sufficient sleep, stress reduction, and cognitive stimulation through mental exercises and learning activities.

Neuroplasticity refers to the brain's ability to change and adapt throughout a person's life. It involves the formation of new neural connections and the reorganization of existing ones. This adaptation can occur in response to various factors, including learning, experience, environmental changes, injury, and aging.

Neuroplasticity plays a significant role in cognitive enhancement because it allows the brain to optimize its functioning and improve cognitive abilities. Here are a few ways in which neuroplasticity contributes to cognitive enhancement:

1. Learning and Memory: Neuroplasticity enables the brain to form new connections and strengthen existing ones, facilitating the acquisition of new knowledge and the consolidation of memories. When we learn something

new, neural pathways are established or modified, allowing for more efficient retrieval of information in the future.

2. Adaptation to Injury: In cases of brain injury or damage, neuroplasticity is crucial for the brain to compensate for the lost functions. The unaffected areas or adjacent regions can undergo changes to assume the functionality of the damaged regions, helping individuals recover or regain cognitive abilities.

3. Training and Skill Development: Through repetitive practice and training, neuroplasticity allows the brain to fine-tune skills and improve performance. With focused practice, neural connections associated with the specific skill strengthen, leading to enhanced motor skills, language proficiency, musical abilities, and more.

4. Cognitive Reserve and Aging: Neuroplasticity is believed to contribute to cognitive reserve, which refers to the brain's ability to withstand age-related changes and neurological disorders without significant cognitive impairment. Engaging in mentally stimulating activities and lifelong learning can help build cognitive reserve and reduce the risk of cognitive decline.

While neuroplasticity occurs naturally, there are various strategies to support and enhance it. These include engaging in mentally challenging activities (such as puzzles, learning a new language, or playing an instrument), incorporating regular physical exercise, getting adequate sleep, managing stress, maintaining a healthy diet, and minimizing harmful habits like excessive alcohol consumption or smoking.

It's worth noting that individual differences exist in terms of baseline neuroplasticity levels and response to interventions. Additionally, the extent of neuroplastic changes and their functional impact can vary depending on age, health, and other factors. Consulting with healthcare professionals or cognitive specialists can provide personalized guidance in leveraging neuroplasticity to enhance cognitive function.

There are several activities that can promote neuroplasticity and support cognitive enhancement. Here are some examples:

1. Learning a New Skill: Engage in activities that involve learning something new and challenging, such as playing a musical instrument, learning a new language, or acquiring a new hobby. These activities stimulate the brain and promote the formation of new neural connections.

2. Cognitive Exercises: Participate in brain-training exercises that target various cognitive functions, including memory, attention, problem-solving, and mental agility. There are numerous apps, online programs, and games specifically designed to enhance cognitive abilities.

3. Physical Exercise: Regular aerobic exercise has been linked to improved brain health and neuroplasticity. Engaging in activities like brisk walking, dancing, swimming, or cycling not only benefits the body but also supports cognitive function.

4. Meditation and Mindfulness: Practicing mindfulness meditation has been shown to have positive effects on neuroplasticity. It can help regulate emotions, reduce stress, and improve attention and mental clarity. Meditation practices like focused attention or loving-kindness meditation can be beneficial.

5. Social Interactions: Engaging in social activities and maintaining strong social connections can contribute to cognitive health. Interacting with others, engaging in conversations, and participating in group activities provide mental stimulation and support brain plasticity.

6. Reading and Mental Stimulation: Reading books, newspapers, or engaging in mentally stimulating activities like puzzles, crosswords, or Sudoku can help keep the brain active and promote neuroplasticity.

7. Healthy Lifestyle: Adopting a healthy lifestyle, including a balanced diet, adequate sleep, and stress management, can support brain health and optimize neuroplasticity.

Remember, consistency and variety are key. It's beneficial to engage in a combination of activities that challenge different cognitive functions and provide ongoing mental stimulation. By incorporating these activities into your routine, you can promote neuroplasticity and support cognitive enhancement.

There are several activities that can promote neuroplasticity and support cognitive enhancement. Here are some examples:

1. Learning a New Skill: Engage in activities that involve learning something new and challenging, such as playing a musical instrument, learning a new language, or acquiring a new hobby. These activities stimulate the brain and promote the formation of new neural connections.

2. Cognitive Exercises: Participate in brain-training exercises that target various cognitive functions, including memory, attention, problem-solving, and mental agility. There are numerous apps, online programs, and games specifically designed to enhance cognitive abilities.

3. Physical Exercise: Regular aerobic exercise has been linked to improved brain health and neuroplasticity. Engaging in activities like brisk walking, dancing, swimming, or cycling not only benefits the body but also supports cognitive function.

4. Meditation and Mindfulness: Practicing mindfulness meditation has been shown to have positive effects on neuroplasticity. It can help regulate emotions, reduce stress, and improve attention and mental clarity. Meditation practices like focused attention or loving-kindness meditation can be beneficial.

5. Social Interactions: Engaging in social activities and maintaining strong social connections can contribute to cognitive health. Interacting with others, engaging in conversations, and participating in group activities provide mental stimulation and support brain plasticity.

6. Reading and Mental Stimulation: Reading books, newspapers, or engaging in mentally stimulating activities like puzzles, crosswords, or Sudoku can help keep the brain active and promote neuroplasticity.

7. Healthy Lifestyle: Adopting a healthy lifestyle, including a balanced diet, adequate sleep, and stress management, can support brain health and optimize neuroplasticity.

Remember, consistency and variety are key. It's beneficial to engage in a combination of activities that challenge different cognitive functions and provide ongoing mental stimulation. By incorporating these activities into your routine, you can promote neuroplasticity and support cognitive enhancement.

Harnessing neurotechnology for therapeutic interventions:

Neurotechnology has revolutionized therapeutic interventions in various medical fields, providing new avenues for treating and managing neurological disorders, chronic pain, and mental health conditions. In this section, we explore some of the key applications of neurotechnology and their impact on patient care.

One of the most promising areas is neurorehabilitation, where neurotechnology is used to aid individuals recovering from neurological disorders or injuries. For example, brain-computer interfaces (BCIs) have shown great potential in assisting patients with motor impairments to regain control of their movements. By translating neural signals into commands that control external devices, BCIs enable patients to perform tasks they might otherwise be unable to accomplish independently.

Another application lies in the realm of mental health. Neurofeedback training, a technique that allows individuals to learn self-regulation of brain activity, has shown promise in treating conditions like ADHD, anxiety, and depression. Using real-time displays of brain activity, individuals can learn to identify and modify their brain patterns, leading to improved emotional regulation and overall well-being.

Additionally, neurostimulation techniques such as transcranial direct current stimulation (tDCS) and deep brain stimulation (DBS) have shown effectiveness in managing conditions like Parkinson's disease and obsessive-compulsive disorder (OCD). TDCS involves the application of low electrical currents to specific areas of the brain, while DBS uses implanted electrodes to deliver controlled electrical impulses. These techniques can alleviate symptoms, improve motor function, and enhance overall quality of life.

Pain management is another area that has greatly benefited from neurotechnology. Chronic pain conditions can be debilitating and challenging to treat, but techniques like spinal cord stimulation (SCS) offer a non-invasive approach to pain relief. By delivering electrical impulses to the spinal cord, SCS disrupts pain signals and provides significant relief to patients, allowing them to engage in daily activities more comfortably.

However, while neurotechnology brings immense potential, there are ethical considerations that need to be addressed. Privacy and data security are paramount, as neurotechnology involves handling sensitive neural data. Protecting patient confidentiality and ensuring proper informed consent are essential for responsible use.

Moreover, access to these neurotechnological interventions should be equitable, without creating further inequalities in healthcare. Ensuring that these advancements are accessible and affordable to all individuals who can benefit from them is crucial for creating an inclusive and just healthcare system.

In conclusion, harnessing neurotechnology for therapeutic interventions has opened up new possibilities for individuals living with neurological disorders, mental health conditions, chronic pain, and more. As research and development continue, we can expect further innovations in neurotechnological interventions, leading to improved patient outcomes and a better quality of life for those in need.

The concept of increasing intelligence is a complex and controversial topic. Intelligence is a multifaceted trait influenced by various genetic, environmental, and developmental factors. At present, there is no widely accepted and verified method to significantly increase overall intelligence in a safe and effective manner.

While there are various brain-training programs and cognitive exercises available, their impact on general intelligence remains a subject of debate among scientists. Some studies suggest that these interventions may lead to improvements in specific cognitive skills but may not necessarily raise overall intelligence levels.

It's also important to note that intelligence is not solely determined by genetics but is influenced by factors such as education, nutrition, and socio-economic conditions. Thus, efforts to enhance intelligence should also focus on providing equal educational opportunities and addressing societal inequalities.

As for the future, it is challenging to predict with certainty when it will be possible to effectively increase intelligence. Ongoing advancements in neuroscience and technology may eventually provide new insights and methods for enhancing cognitive abilities. However, ethical considerations, potential side effects, and the need for rigorous scientific validation would need to be carefully addressed before any cognitive enhancement interventions can be widely implemented.

Ultimately, the goal should be to foster a society that values and supports lifelong learning, critical thinking, and intellectual development. By promoting access to quality education, encouraging personal growth, and creating an environment that prioritizes intellectual pursuits, we can contribute to the creation of a more intellectually engaged and informed citizenry.

Promoting intellectual development in society is crucial for personal growth, innovation, and societal progress. Here are some strategies to foster intellectual development:

1. Emphasize and prioritize education: Ensure access to quality education for all individuals, regardless of socio-economic background. Improve educational resources, infrastructure, and teaching methodologies to create an environment conducive to intellectual growth.

2. Encourage critical thinking: Teach individuals to question and analyze information critically. Encourage open-mindedness, evidence-based reasoning, and the ability to evaluate different perspectives and viewpoints.

3. Promote lifelong learning: Emphasize the importance of continuous learning and personal growth throughout life. Encourage individuals to explore new subjects, acquire new skills, and engage in intellectual pursuits even beyond formal education.

4. Support intellectual curiosity: Encourage individuals to explore their interests, pursue their passions, and engage in intellectual activities such as reading, writing, and research. Provide spaces and resources that facilitate intellectual exploration and creativity.

5. Foster a culture of intellectual discourse: Encourage respectful and inclusive debates, discussions, and exchange of ideas. Create platforms where individuals can share their thoughts and opinions, stimulating intellectual growth through meaningful dialogue.

6. Provide opportunities for intellectual engagement: Offer programs, workshops, and extracurricular activities that promote intellectual development. Encourage participation in activities like debates, research projects, scientific experiments, literary clubs, and art appreciation to cultivate diverse intellectual interests.

7. Support information literacy: Teach individuals how to navigate and critically evaluate information in today's information-rich society. Promote media literacy skills to enable individuals to distinguish between credible sources and misinformation.

8. Create supportive environments: Establish environments that value intellectual pursuits, innovation, and intellectual diversity. Encourage collaboration, mentorship, and the sharing of knowledge and ideas.

9. Lead by example: Role models, such as educators, parents, and community leaders, play a crucial role in promoting intellectual development. Demonstrating a passion for learning, intellectual curiosity, and respect for intellectual achievements can inspire others to follow suit.

Overall, creating a society that values and promotes intellectual development requires a comprehensive approach that encompasses education, access to resources, cultural attitudes, and supportive environments. By implementing these strategies, we can foster a society where individuals are encouraged to develop their intellectual potential to contribute positively to their communities and the world.

Cultivating well-being and finding a balance between physical and mental well-being is essential for leading a healthy and fulfilling life. Here are some strategies to help achieve this balance:

1. Prioritize self-care: Take care of your physical and mental health by engaging in activities that promote well-being. This might include regular exercise, sufficient sleep, nutritious eating, practicing mindfulness or meditation, and engaging in hobbies or activities that bring you joy.

2. Maintain a balanced lifestyle: Strive for a well-rounded lifestyle that includes time for work, rest, recreation, and social connections. Avoid excessive work or personal commitments that can lead to burnout and neglecting other aspects of your life.

3. Practice stress management: Learn and implement stress management techniques such as deep breathing exercises, journaling, or engaging in activities that help you relax and unwind. Find healthy ways to cope with stress and build resilience.

4. Nurture social connections: Cultivate meaningful relationships and social connections with friends, family, and community. Engage in activities that promote social interactions and support systems, as they contribute to both physical and mental well-being.

5. Seek professional help if needed: If you're struggling with your mental well-being, don't hesitate to seek help from mental health professionals. They can provide guidance, support, and therapeutic interventions to help you navigate challenges and enhance your overall well-being.

6. Practice mindful technology use: While technology can be helpful, excessive use and dependency on screens can negatively impact both physical and mental well-being. Set boundaries for screen time and cultivate a healthy relationship with technology to maintain balance.

7. Engage in physical activity: Regular exercise not only benefits physical health but also has a positive impact on mental well-being. Find activities you enjoy, whether it's going for walks, practicing yoga, or playing a sport. Incorporate movement into your daily routine.

8. Take breaks and recharge: Recognize the importance of rest and relaxation. Make time for breaks throughout the day and plan regular vacations or time off to recharge and rejuvenate.

9. Practice gratitude and positive mindset: Cultivate a mindset of gratitude and focus on the positive aspects of life. Regularly express gratitude for the things you appreciate, and adopt optimistic thinking patterns to enhance mental well-being.

Remember that finding balance and well-being is an ongoing process. It's important to listen to your body and mind, and make adjustments as necessary to maintain a healthy equilibrium between physical and mental well-being.

The impact of technological singularity on the job market is a topic of significant concern and debate. As AI and automation technologies continue to advance, there is a potential for widespread job displacement and transformation across various industries.

On one hand, the increased automation of repetitive tasks and the emergence of AI-powered systems can lead to increased productivity and efficiency in many sectors. This may result in the creation of new job roles, as well as the need for workers with skills that complement and collaborate with AI technologies.

However, there is also a legitimate concern that certain jobs could become obsolete as AI systems become more capable of performing complex tasks. Jobs that are easily automated, such as those involving routine manual or cognitive tasks, are particularly at risk. This includes positions in manufacturing, transportation, data entry, and even some aspects of healthcare and professional services.

The displacement of workers due to automation can lead to economic and social challenges, including job losses, income inequality, and skills gaps. It is crucial for societies and governments to proactively respond to these challenges by investing in retraining and reskilling programs, supporting affected workers in transition, and creating new job opportunities in emerging fields.

Furthermore, it is essential to recognize that while AI may take over certain job functions, it can also create new opportunities for human workers. AI can enhance cognitive capabilities, provide support for decision-making, and augment human skills. Therefore, a focus on developing skills that AI cannot easily replicate, such as creativity, problem-solving, emotional intelligence, and interpersonal communication, becomes increasingly important.

In summary, the impact of technological singularity on the job market is complex and multifaceted. While there is potential for job displacement, there are also opportunities for new job creation and the augmentation of human capabilities. To mitigate the negative impacts, it is crucial for governments, industries, and individuals to embrace lifelong learning, invest in reskilling, and ensure that the benefits of AI and automation are shared equitably across society.

Several industries are more likely to be affected by automation as AI and advanced technologies continue to advance. While the extent of automation can vary within each industry, here are some sectors that are commonly seen as having a higher potential for automation impact:

1. Manufacturing and production: Automation has already revolutionized manufacturing processes, with robotics and AI-enabled systems handling tasks like assembly, quality control, and logistics.

2. Transportation and logistics: The rise of self-driving vehicles and advancements in smart logistics systems have the potential to automate functions such as shipping, delivery, and warehouse management.

3. Retail: AI-powered chatbots, self-checkout systems, and automated warehouses are examples of automation technologies being adopted in the retail sector to streamline operations and enhance customer experiences.

4. Healthcare: While many healthcare roles require human expertise, automation is being explored in areas such as diagnostics, administrative tasks, and robotic-assisted surgeries.

5. Customer service: AI-powered chatbots and virtual assistants are increasingly being used in customer support roles to handle common queries and provide assistance, reducing the need for human interaction.

6. Data entry and administrative tasks: Many repetitive and rule-based administrative tasks, such as data entry and paperwork, can be automated using AI-powered systems.

7. Finance and banking: Automation is already impacting financial and banking services with the use of AI-based software for tasks like fraud detection, risk assessment, and algorithmic trading.

8. Agriculture: Automation technologies like precision farming, robotic harvesting, and drone monitoring are transforming the agricultural sector, increasing productivity and efficiency.

9. Construction: Robots and automated machinery can assist in various construction tasks, such as bricklaying, concrete pouring, and even inspection.

10. Legal services: AI technologies are being utilized in areas such as legal research, contract analysis, and document review to streamline processes and increase efficiency.

It is important to note that while these industries may see increased automation, it does not necessarily mean that all jobs within these sectors will be completely replaced. Automation is more likely to impact specific tasks and job functions, leading to the need for upskilling, reskilling, and the creation of new roles that can work alongside AI technologies.

Chapter 13: Overcoming Technological Singularity

A. Understanding the concept of technological singularity:

Technological singularity refers to a hypothetical point in the future when artificial intelligence (AI) becomes capable of surpassing human intelligence. It is a point where AI systems can improve themselves autonomously, leading to an exponential growth in their capabilities. Understanding the concept of technological singularity is important because it has significant implications for society, ethics, and the future of humanity.

B. Preparing for the potential challenges and opportunities:

As we approach the potential realization of technological singularity, it is crucial to prepare for both the challenges and opportunities that it presents. On the one hand, the rapid advancement of AI and superintelligence could bring about breakthroughs in medicine, energy, and scientific discoveries. On the other hand, there are concerns about job displacement, ethical dilemmas, and potential dangers if AI is not properly controlled or aligned with human values.

To prepare for these challenges and opportunities, it is essential to invest in research and development, foster interdisciplinary collaborations, and engage in ongoing discussions and debates. Society needs to anticipate and address the ethical, social, and economic impacts of AI and superintelligence, exploring potential regulations, policy frameworks, and safeguards.

C. Building robust safeguards and ethical frameworks:

As we navigate the implications of technological singularity, it is important to establish robust safeguards and ethical frameworks. This involves creating guidelines, standards, and regulations to ensure that AI development and deployment align with human values, prioritize safety, and uphold fundamental rights and ethics.

One crucial aspect is to implement transparency and accountability mechanisms, ensuring that AI systems are explainable, auditable, and accountable for their decisions and actions. Collaborative efforts between researchers, policymakers, and industry stakeholders are essential to establish norms and standards that govern AI development, deployment, and use.

Additionally, frameworks for responsible AI should address biases, fairness, and privacy concerns. They should also consider the potential impact on employment and work structures, fostering inclusive approaches that promote human well-being and social equity.

Building ethical frameworks also requires ongoing dialogue and engagement with diverse stakeholders, including experts in various disciplines, public representatives, and affected communities. By incorporating a wide range of perspectives, we can collectively shape the ethical development and deployment of AI technologies, ensuring they benefit society as a whole.

In conclusion, understanding technological singularity, preparing for its potential challenges and opportunities, and building robust safeguards and ethical frameworks are crucial steps as we move forward. By staying informed, proactive, and committed to ethical considerations, we can effectively navigate the path towards a future where advanced technologies, including artificial intelligence, enhance human well-being and contribute to a more equitable and sustainable society.

The impact of technological singularity on the job market is a topic of significant concern and debate. As AI and automation technologies continue to advance, there is a potential for widespread job displacement and transformation across various industries.

On one hand, the increased automation of repetitive tasks and the emergence of AI-powered systems can lead to increased productivity and efficiency in many sectors. This may result in the creation of new job roles, as well as the need for workers with skills that complement and collaborate with AI technologies.

However, there is also a legitimate concern that certain jobs could become obsolete as AI systems become more capable of performing complex tasks. Jobs that are easily automated, such as those involving routine manual or cognitive tasks, are particularly at risk. This includes positions in manufacturing, transportation, data entry, and even some aspects of healthcare and professional services.

The displacement of workers due to automation can lead to economic and social challenges, including job losses, income inequality, and skills gaps. It is crucial for societies and governments to proactively respond to these challenges by investing in retraining and reskilling programs, supporting affected workers in transition, and creating new job opportunities in emerging fields.

Furthermore, it is essential to recognize that while AI may take over certain job functions, it can also create new opportunities for human workers. AI can enhance cognitive capabilities, provide support for decision-making, and augment human skills. Therefore, a focus on developing skills that AI cannot easily replicate, such as creativity, problem-solving, emotional intelligence, and interpersonal communication, becomes increasingly important.

In summary, the impact of technological singularity on the job market is complex and multifaceted. While there is potential for job displacement, there are also opportunities for new job creation and the augmentation of human capabilities. To mitigate the negative impacts, it is crucial for governments, industries, and individuals to embrace lifelong learning, invest in reskilling, and ensure that the benefits of AI and automation are shared equitably across society.

Several industries are more likely to be affected by automation as AI and advanced technologies continue to advance. While the extent of automation can vary within each industry, here are some sectors that are commonly seen as having a higher potential for automation impact:

1. Manufacturing and production: Automation has already revolutionized manufacturing processes, with robotics and AI-enabled systems handling tasks like assembly, quality control, and logistics.

2. Transportation and logistics: The rise of self-driving vehicles and advancements in smart logistics systems have the potential to automate functions such as shipping, delivery, and warehouse management.

3. Retail: AI-powered chatbots, self-checkout systems, and automated warehouses are examples of automation technologies being adopted in the retail sector to streamline operations and enhance customer experiences.

4. Healthcare: While many healthcare roles require human expertise, automation is being explored in areas such as diagnostics, administrative tasks, and robotic-assisted surgeries.

5. Customer service: AI-powered chatbots and virtual assistants are increasingly being used in customer support roles to handle common queries and provide assistance, reducing the need for human interaction.

6. Data entry and administrative tasks: Many repetitive and rule-based administrative tasks, such as data entry and paperwork, can be automated using AI-powered systems.

7. Finance and banking: Automation is already impacting financial and banking services with the use of AI-based software for tasks like fraud detection, risk assessment, and algorithmic trading.

8. Agriculture: Automation technologies like precision farming, robotic harvesting, and drone monitoring are transforming the agricultural sector, increasing productivity and efficiency.

9. Construction: Robots and automated machinery can assist in various construction tasks, such as bricklaying, concrete pouring, and even inspection.

10. Legal services: AI technologies are being utilized in areas such as legal research, contract analysis, and document review to streamline processes and increase efficiency.

It is important to note that while these industries may see increased automation, it does not necessarily mean that all jobs within these sectors will be completely replaced. Automation is more likely to impact specific tasks and job functions, leading to the need for upskilling, reskilling, and the creation of new roles that can work alongside AI technologies.

B. Preparing for the potential challenges

Here are some ways to prepare for the potential challenges that may arise from the impact of automation on the job market:

1. Lifelong learning: Embrace the concept of continuous learning and upskilling to adapt to changing job requirements and new technologies. Stay updated on industry trends and acquire new skills that complement automation, such as data analysis, programming, or creative problem-solving.

2. Focus on uniquely human skills: Develop and enhance skills that are difficult to automate, such as creativity, critical thinking, complex problem-solving, emotional intelligence, and interpersonal communication. These skills can often complement the capabilities of automation and AI, making you more valuable in the workforce.

3. Adaptability and flexibility: Be open to career transitions and new job opportunities that may arise as automation changes the landscape of different industries. Stay agile and identify areas where your skills can be transferred or augmented by new technologies.

4. Reskilling and job transition support: Governments, organizations, and educational institutions should invest in reskilling programs to support workers who may face dislocation due to automation. This can involve providing training in emerging fields, offering financial assistance for education or retraining, and ensuring job placement support.

5. Collaboration between humans and machines: Embrace the concept of human-machine collaboration. Rather than seeing automation as a complete replacement for jobs, focus on how AI and automation technologies can augment human capabilities and assist with more complex tasks. This approach can lead to the creation of new roles and opportunities.

6. Social safety nets and income support: Governments and policymakers need to consider the potential impact of automation on jobs and ensure that appropriate social safety nets are in place to support individuals who face job displacement. This may involve providing unemployment benefits, income support, and creating policies that promote fair distribution of the benefits of automation.

7. Ethical considerations: As automation technologies advance, there will be important ethical considerations to address. This includes addressing biases in algorithms, ensuring transparency in decision-making systems, and preserving human values and ethical standards in the development and use of AI technologies.

By actively preparing for the potential challenges, both individuals and societies can proactively navigate the evolving job market and harness the benefits of automation while minimizing its negative impacts.

C. Building robust safeguards and ethical frameworks

Building robust safeguards and ethical frameworks is crucial in ensuring the responsible development and deployment of automation technologies. Here are some key steps to consider:

1. Ethical guidelines and regulations: Governments, industry associations, and organizations should work together to establish clear ethical guidelines and regulations for the development and use of automation technologies. These guidelines should address issues such as privacy, security, fairness, transparency, and accountability.

2. Bias detection and mitigation: Measures should be put in place to identify and eliminate biases in automated systems. Developers should regularly review and test algorithms for potential biases, ensuring that

automated processes do not discriminate against certain individuals or groups.

3. Data privacy and security: With automation relying on vast amounts of data, it is important to implement robust data privacy and security measures. Strict protocols need to be in place to protect sensitive personal and corporate data from unauthorized access, breaches, or misuse.

4. Human oversight and decision-making: Critical decision-making processes should involve human oversight and intervention. Even with automation, there should be checks and balances to ensure human accountability and the ability to intervene if necessary.

5. Transparency and explainability: Employ transparency in automated systems, making it clear to users how decisions are being made and what data is being used. Explainability is crucial to build trust and understanding in automated processes, especially in areas like AI-based decision-making algorithms.

6. Continuous monitoring and evaluation: Regular monitoring and evaluation of automated systems are necessary to identify any potential risks or biases that may emerge over time. This enables prompt corrective measures and improvements to be implemented.

7. Engaging diverse perspectives: Include diverse perspectives in the development and decision-making process surrounding automation. This can help identify potential biases and ensure that automation technologies are inclusive, fair, and beneficial for all.

8. Public awareness and education: Engage in public awareness campaigns and educational initiatives to inform the public about automation technologies, their impact, and the associated ethical considerations. Empowering individuals with knowledge helps foster informed discussions and decision-making.

9. International collaboration: Encourage international collaboration and cooperation in establishing ethical frameworks and standards for automation technologies. These issues transcend national boundaries, and global cooperation is essential in addressing shared challenges.

Building robust safeguards and ethical frameworks is a continuous process that requires ongoing dialogue and adaptation to keep pace with technological advancements. By prioritizing ethics and responsible development, we can ensure that automation technologies contribute positively to society while minimizing potential risks and challenges.

Chapter 14: Collaboration at Global Scale

A. Fostering international cooperation for global challenges

Fostering international cooperation is critical for addressing global challenges collaboratively. Here are some important steps to promote and enhance collaboration at a global scale:

1. Multilateral organizations and agreements: Strengthen and support existing multilateral organizations such as the United Nations, World Health Organization, World Trade Organization, and International Monetary Fund. These organizations play a crucial role in facilitating cooperation and coordinating efforts to tackle global challenges.

2. Diplomatic engagement: Promote diplomatic engagement and communication among nations to foster mutual understanding, trust, and cooperation. Regular dialogues, exchanges, and diplomacy can help build relationships and facilitate collaboration on various issues.

3. Sharing knowledge and best practices: Encourage the sharing of knowledge, research findings, and best practices among countries. This can be achieved through international conferences, forums, collaborations between research institutions, and open access to scientific publications.

4. Financial assistance and resource sharing: Support international financial mechanisms to provide assistance to countries facing challenges. This can include aid programs, debt relief initiatives, and funding for infrastructure development, healthcare, education, and sustainable development.

5. Technology transfer and capacity-building: Facilitate technology transfer from developed nations to developing nations, helping them build their capabilities and infrastructure to address global challenges. This can involve sharing technical expertise, providing training programs, and supporting technology initiatives that benefit all nations.

6. Collective action on climate change: Strengthen cooperation and collective action on climate change mitigation and adaptation. Encourage nations to commit to and implement the Paris Agreement, encouraging the

transition to sustainable energy sources, reducing greenhouse gas emissions, and protecting vulnerable ecosystems.

7. Public-private partnerships: Foster collaboration between governments, private sector organizations, and civil society to leverage their respective strengths in addressing global challenges. Public-private partnerships can bring together resources, expertise, and innovation to tackle complex issues more effectively.

8. Crisis response coordination: Enhance coordination and response mechanisms for global crises such as pandemics, natural disasters, and humanitarian emergencies. This can involve establishing early warning systems, sharing disease surveillance data, and coordinating relief efforts to ensure a swift and efficient response.

9. Cultural exchange and people-to-people connections: Promote cultural exchange programs, student exchanges, and people-to-people connections to foster global understanding and cooperation. Engaging with different cultures and perspectives can break down barriers, build empathy, and facilitate collaboration.

10. Sustainable development goals: Align efforts with the United Nations Sustainable Development Goals (SDGs) to address global challenges comprehensively. Collaborate on initiatives related to poverty eradication, education, gender equality, clean energy, sustainable cities, and responsible consumption and production.

By fostering international cooperation for global challenges, we can pool resources, knowledge, and expertise to create a more inclusive and sustainable future for all nations. Collaboration allows us to tackle shared problems collectively, find innovative solutions, and build a better world together.

B. Overcoming cultural, economic, and political barriers

Overcoming cultural, economic, and political barriers is essential to fostering effective collaboration at a global scale. Here are some strategies to address these barriers:

1. Cultural sensitivity and understanding: Cultivate cultural sensitivity and awareness to navigate cultural differences. Encourage open dialogue, active listening, and respect for diverse cultural perspectives and practices. This

can help build trust, foster understanding, and overcome cultural barriers that may hinder collaboration.

2. Language and communication: Facilitate effective communication by ensuring language accessibility and promoting language learning. The use of interpreters, translation services, and multilingual resources can aid in bridging linguistic barriers. Emphasize clear and inclusive communication to prevent misunderstandings and promote effective collaboration.

3. Economic cooperation and inclusive growth: Promote fair economic cooperation that benefits all parties involved. Foster trade relations, investment partnerships, and knowledge-sharing to bridge economic disparities and promote inclusive growth. Addressing economic barriers can lead to mutually beneficial collaborations and sustainable development.

4. Political dialogue and diplomacy: Enhance political dialogue and diplomacy to address political barriers that hinder collaboration. Foster a culture of cooperation, negotiation, and compromise among nations to find common ground on global issues. Strengthen diplomatic channels, multilateral forums, and international agreements to facilitate political collaboration.

5. Peacebuilding and conflict resolution: Address conflicts and work towards peaceful resolutions to overcome political and social barriers. Promote dialogue, mediation, and reconciliation to build trust and promote stability. Resolving conflicts opens up opportunities for collaboration on shared challenges.

6. Inclusive decision-making: Ensure inclusivity in decision-making processes to overcome power imbalances and political barriers. Emphasize inclusive and participatory approaches that involve various stakeholders, including marginalized groups, civil society organizations, and local communities. This creates a sense of ownership and fosters collaboration.

7. Cultural exchange and cross-cultural training: Encourage cultural exchange programs, cross-cultural training, and experiential learning opportunities. This allows individuals and organizations to develop cross-cultural competence and understand different societal norms, values, and perspectives. Increased cultural understanding helps bridge gaps and promotes collaboration.

8. Economic incentives and financial support: Provide economic incentives and financial support for collaboration initiatives. This can include grants,

funding opportunities, tax incentives, and rewards for cross-border collaborations. Financial support can help overcome economic barriers and motivate collaboration between different nations.

9. Building trust and relationships: Focus on building trust and nurturing strong relationships among stakeholders. Encourage regular interactions, collaboration projects, and joint initiatives to foster trust and overcome barriers caused by lack of trust or previous conflicts.

10. Advocacy for inclusivity and collaboration: Promote the importance of collaboration, inclusivity, and overcoming barriers at various levels. Advocate for policies, programs, and initiatives that prioritize collaboration at the global scale. This can help create a supportive environment that encourages collaboration and reduces cultural, economic, and political barriers.

By actively addressing and overcoming cultural, economic, and political barriers, we can create an environment that fosters effective collaboration and enables global solutions for the challenges we face.

There are several examples of economic cooperation that promote collaboration and address economic barriers. Here are a few notable ones:

1. Free Trade Agreements (FTAs): FTAs, such as the North American Free Trade Agreement (NAFTA) and the European Union (EU), promote economic cooperation by eliminating or reducing trade barriers such as tariffs, quotas, and import/export restrictions. These agreements facilitate the flow of goods and services, encourage investment, and stimulate economic growth among participating countries.

2. Economic and Monetary Unions: Economic and monetary unions, like the Eurozone, establish a common currency and economic policies among member countries. This fosters economic cooperation, facilitates cross-border trade and investment, and creates a unified economic market within the union.

3. Bilateral Investment Treaties (BITs): BITs are agreements between two countries that promote and protect foreign investment. These treaties provide legal frameworks that support economic cooperation by facilitating investment flows, protecting investors' rights, and resolving investment disputes.

4. Development Assistance and Aid: Economic cooperation can also be seen through development assistance and aid programs. Developed countries provide financial and technical support to developing nations to promote economic growth, alleviate poverty, and build sustainable infrastructure. These programs aim to foster economic cooperation and reduce economic disparities.

5. Regional Economic Integration Organizations: Regional economic integration organizations, such as the Association of Southeast Asian Nations (ASEAN) and the Mercosur in South America, promote economic cooperation among member countries. They focus on harmonizing trade policies, facilitating regional trade, and enhancing economic coordination for shared prosperity.

6. Cross-border Infrastructure Projects: Collaborative infrastructure projects, like the Belt and Road Initiative (BRI) led by China, promote economic cooperation by enhancing connectivity and trade routes among participating nations. These projects aim to improve transportation, energy, and digital infrastructure, facilitating economic integration and cooperation.

7. Joint Research and Development (R&D) Initiatives: Economic cooperation can occur through joint R&D initiatives, where countries or companies pool resources, knowledge, and expertise to develop new technologies, products, and solutions. By collaborating on research and development, participating entities can leverage each other's strengths, share costs, and accelerate innovation.

8. Supply Chains and Global Value Chains (GVCs): Supply chains and GVCs involve the interconnected network of activities, production, and trade across different countries. Economic cooperation is evident in the sharing of resources, technology, and expertise among participating nations, contributing to increased efficiency and competitiveness.

These examples demonstrate how economic cooperation can foster collaboration, create economic interdependence, and promote shared growth among nations.

Economic cooperation in Europe is quite significant and takes various forms. Here are a few examples:

1. European Union (EU): The European Union is a prime example of economic cooperation in Europe. It consists of 27 member countries who have come together to promote peace, stability, and economic growth through a common market. The EU facilitates the free movement of goods, services, capital, and labor among member states, eliminating trade barriers and promoting economic integration.

2. Eurozone: The Eurozone is a subgroup within the EU that comprises 19 member countries that have adopted the euro as their common currency. This monetary union enhances economic cooperation by enabling seamless transactions and trade within the Eurozone, fostering price stability and monetary policy coordination.

3. Single Market: The EU single market is a core element of economic cooperation in Europe. It ensures the free movement of goods, services, capital, and people within the EU. By harmonizing regulations, standardizing product requirements, and removing trade barriers, the single market facilitates cross-border trade and stimulates economic activity.

4. Customs Union: The EU also operates as a customs union, which means that member countries apply a common external tariff and have a unified trade policy with non-EU countries. This encourages economic cooperation by eliminating tariffs and other trade barriers between member states while presenting a united economic front to the rest of the world.

5. European Economic Area (EEA): The EEA consists of the EU member states as well as Iceland, Liechtenstein, and Norway. It facilitates economic cooperation by extending the EU's single market principles to these non-EU countries, allowing them to participate in the free movement of goods, services, capital, and people within the EEA.

6. Regional Development Funds: The EU offers regional development funds to support economic cooperation and reduce economic disparities among its member countries. These funds provide financial assistance for infrastructure projects, research and innovation, and regional development initiatives, promoting economic cohesion and cooperation across Europe.

7. Cross-Border Cooperation: Numerous cross-border cooperation initiatives exist within Europe, focusing on enhancing economic ties and cooperation between neighboring regions in different countries. These initiatives promote joint investments, infrastructure development, and collaboration in areas such as tourism, trade, and innovation, fostering economic integration at a local level.

These examples illustrate the depth and breadth of economic cooperation in Europe, highlighting the EU's efforts to strengthen economic ties, facilitate trade, and promote shared prosperity among its member countries.

C. Uniting humanity toward shared goals and aspirations.

Uniting humanity toward shared goals and aspirations is indeed an admirable objective, and AI can play a vital role in facilitating collaborative efforts and harnessing collective intelligence. Here are a few ways in which AI can contribute to this goal:

1. Enhancing Communication: AI technologies can bridge language barriers by providing real-time translation and interpretation services. This can facilitate effective communication and understanding among people from different cultures and languages, enabling them to work together towards shared goals more seamlessly.

2. Promoting Knowledge Sharing: AI-powered platforms and tools can enable the sharing of knowledge and expertise on a global scale. Through online communities, forums, and collaborative platforms, individuals from diverse backgrounds can exchange insights, ideas, and solutions, fostering innovation and cooperation in pursuit of common aspirations.

3. Identifying Global Challenges: AI can help identify and analyze complex global challenges, such as climate change, poverty, and public health crises. By analyzing vast amounts of data and providing insights, AI systems can support informed decision-making and facilitate collective understanding of these challenges, encouraging global collaboration to find solutions.

4. Assisting in Decision-Making: AI algorithms can aid in decision-making processes by providing objective analyses and predictions. By incorporating diverse perspectives and considering a wide range of factors, AI-powered decision support systems can promote fairness, inclusivity, and collective decision-making based on shared goals and aspirations.

5. Facilitating Global Cooperation: AI can facilitate cooperation among individuals and organizations across borders. For instance, AI-powered virtual collaboration tools can enable remote teamwork, allowing individuals from different parts of the world to collaborate effectively on shared projects and initiatives.

6. Personalizing Experiences: AI algorithms can personalize experiences for individuals by understanding their preferences, needs, and aspirations. By tailoring information, recommendations, and opportunities to individuals' interests, AI can connect people with like-minded individuals, organizations, and resources, fostering connections and collaborations for shared goals.

However, it is important to be mindful of the ethical considerations and potential biases associated with AI technologies. To ensure that AI contributes positively to our shared goals and aspirations, it is crucial to develop and deploy AI systems in a transparent, accountable, and inclusive manner, while addressing privacy concerns and potential biases.

AI, when used responsibly and ethically, has the potential to be a powerful tool in uniting humanity towards shared goals and aspirations, promoting collaboration, inclusivity, and mutual understanding on a global scale.

AI can promote knowledge sharing in several ways:

1. Content Recommendations: AI-powered algorithms can analyze users' preferences, search history, and behavior patterns to provide personalized content recommendations. These recommendations can help individuals discover relevant information, articles, research papers, and resources that align with their interests, encouraging knowledge exploration and sharing.

2. Collaborative Filtering: AI can employ collaborative filtering techniques to connect users with similar interests or expertise. By analyzing the behavior and preferences of different users, AI algorithms can identify patterns and recommend relevant content or connect people with similar interests, fostering knowledge exchange and collaboration.

3. Natural Language Processing: AI can utilize natural language processing techniques to analyze and extract valuable insights from vast amounts of text-based data, including articles, research papers, and online forums. By automatically summarizing and categorizing information, AI can provide concise and easily accessible knowledge that encourages sharing.

4. Question-Answering Systems: AI-powered question-answering systems, such as chatbots or virtual assistants, can provide instant and accurate responses to inquiries. These systems leverage natural language processing and machine learning algorithms to understand and respond to users' questions, facilitating the sharing of knowledge and information in real-time.

5. Online Collaboration Tools: AI can enhance online collaboration by providing intelligent features in collaborative platforms. For example, AI algorithms can suggest relevant documents, resources, or experts based on the context of a conversation. This can optimize the knowledge-sharing process by connecting individuals with the right information and people at the right time.

6. Semantic Analysis: AI can apply semantic analysis techniques to understand the underlying meaning and context of written or spoken content. By recognizing patterns, relationships, and concepts within the text, AI can facilitate better content organization and relevance, making knowledge sharing more effective and efficient.

By leveraging these AI-powered approaches, individuals and organizations can benefit from improved access to relevant information, personalized learning experiences, and enhanced collaboration opportunities, ultimately promoting knowledge sharing on a global scale.

Natural language processing and AI.

Natural Language Processing (NLP) is a branch of AI that focuses on the interaction between computers and human language. It involves the development of algorithms and models to understand, interpret, and generate human language in a way that is both meaningful and useful.

Here are some key aspects and techniques used in NLP:

1. Text Preprocessing: Before analyzing and processing textual data, NLP often involves various preprocessing steps, such as tokenization, stemming, and lemmatization. Tokenization involves breaking text into individual words or tokens, stemming reduces words to their base or root form, and lemmatization converts words to their dictionary form.

2. Named Entity Recognition (NER): NER is a technique in which NLP models identify and classify named entities in text. These entities can include names of people, organizations, locations, dates, and other specific terms. NER helps extract important information from unstructured text and can be useful in various applications such as information retrieval and question answering.

3. Word Embeddings: Word embeddings are numerical vector representations of words that capture semantic relationships between words. Embeddings allow NLP models to understand the meaning of words based on their contextual usage. Popular word embedding models include Word2Vec, GloVe, and FastText.

4. Sentiment Analysis: Sentiment analysis, also known as opinion mining, involves determining the sentiment or subjective tone of a piece of text. NLP models can analyze text and classify it as positive, negative, or neutral, allowing businesses and organizations to gauge public opinion from social media posts, customer reviews, and other sources.

5. Machine Translation: NLP models can be used for machine translation, translating text from one language to another. Techniques such as statistical machine translation and neural machine translation utilize large bilingual corpora and language models to generate accurate translations.

6. Text Summarization: NLP models can automatically generate concise summaries of longer texts, including articles, documents, or news articles. Extractive summarization methods select important sentences directly from the original text, while abstractive summarization techniques generate new sentences that capture the essence of the original content.

7. Question-Answering Systems: NLP models can be trained to understand and answer questions posed in natural language. These systems interpret the question, search for relevant information, and generate an accurate response. Question-answering systems can be applied in chatbots, virtual assistants, or information retrieval systems.

These are just a few examples of the wide array of techniques and applications within natural language processing. NLP plays a crucial role in enabling computers to understand and process human language, unlocking various AI-powered capabilities for language understanding, information extraction, and communication.

Chapter 15: The Role of Education

A. Rethinking education for the future

Rethinking education for the future:

In today's rapidly evolving world, traditional educational models face the challenge of preparing students for jobs and opportunities that may not even exist yet. Therefore, it is crucial to rethink education and adapt it to the demands of the future. Here are some key areas to consider:

1. Emphasizing critical thinking and problem-solving: Traditional education often focuses heavily on memorization and regurgitation of information. However, in the future, the ability to think critically, analyze complex problems, and find innovative solutions will be increasingly valuable. Education should encourage students to develop these skills through project-based learning, case studies, and real-world problem-solving activities.

2. Technology integration: The advancement of technology presents both challenges and opportunities for education. Integrating appropriate technology tools and platforms in the classroom can help enhance the learning experience, engage students, and prepare them for the digital world. This includes leveraging AI-driven tools, virtual reality, augmented reality, and online collaboration platforms to create interactive and immersive learning environments.

3. Encouraging adaptability and lifelong learning: With the pace of change accelerating, it is essential to instill a growth mindset in students, fostering a love of learning and a willingness to adapt to new challenges. Education should emphasize the development of skills such as resilience, flexibility, communication, and continuous learning. By equipping students with these skills, they will be better prepared to navigate an ever-changing job market.

4. Promoting interdisciplinary learning: The complex problems of the future often require a multidisciplinary approach. Education should encourage students to explore diverse subjects and make connections across different disciplines. By integrating knowledge from various fields, students can develop a holistic understanding of the world and gain the ability to solve complex problems from multiple perspectives.

5. Developing social and emotional skills: In addition to academic knowledge, education should focus on nurturing social and emotional skills. This includes promoting empathy, collaboration, effective communication, and cultural awareness. These skills are vital for students to thrive in diverse work environments and contribute positively to society.

6. Fostering creativity and innovation: The ability to think creatively, generate new ideas, and innovate will be essential for success in the future. Education should provide opportunities for students to explore their

creativity, engage in creative problem-solving, and develop their entrepreneurial skills. This can be fostered through the integration of arts, design thinking methodologies, and entrepreneurship programs.

7. Personalized and adaptive learning: Recognizing that every student is unique and has different learning needs, education should strive to provide personalized and adaptive learning experiences. AI and data analytics can be leveraged to understand individual learning styles, preferences, and progress, allowing for tailored instruction and personalized feedback.

By embracing these principles and rethinking education for the future, we can better equip students with the skills and mindset needed to thrive in an uncertain and dynamic world. Education should aim to empower individuals to become lifelong learners, critical thinkers, problem solvers, and contributors to a more inclusive and innovative society.

Integrating technology into traditional classrooms can enhance the learning experience and provide students with new opportunities for engagement and knowledge acquisition. Here are some ways to do so:

1. Digital resources and multimedia content: Teachers can incorporate digital resources like e-books, educational apps, interactive websites, and online videos into their lessons. These resources can supplement and reinforce classroom instruction, providing students with a more dynamic and interactive learning experience.

2. Interactive whiteboards and projectors: Utilizing interactive whiteboards or projectors allows teachers to present information in a visually engaging manner. They can display multimedia content, interactive activities, and live demonstrations, making lessons more captivating for students.

3. Online collaboration tools: Virtual collaboration tools, such as Google Docs, Microsoft Teams, or other platforms, enable students to collaborate on projects, share documents, and work together remotely. These tools promote teamwork, communication, and the development of digital literacy skills.

4. Online assessments and feedback: Technology can streamline the assessment process by automating grading and providing instant feedback to students. Online quizzes, tests, and assignments allow educators to provide timely feedback and identify areas for improvement, helping students learn from their mistakes and progress more effectively.

5. **Virtual reality (VR) and augmented reality (AR):** VR and AR technologies can be used to create immersive and interactive learning experiences. Students can explore historical landmarks, visit virtual museums, or participate in virtual simulations to enhance their understanding of complex concepts, making learning more engaging and memorable.

6. **Online learning platforms:** Learning management systems (LMS) like Canvas, Moodle, or Schoology can be used to deliver course materials, assignments, and quizzes online. These platforms provide a centralized hub where students can access resources, submit work, and interact with their peers and teachers outside of class time.

7. **Educational apps and gamification:** Educational apps and gamified learning platforms make learning fun and interactive. They often incorporate game elements, rewards, and challenges, motivating students to actively participate and reinforcing their learning in an enjoyable way.

8. **Bring Your Own Device (BYOD) policies:** Schools can encourage students to bring their own devices, such as laptops, tablets, or smartphones, to utilize in the classroom. This allows for personalized learning experiences and facilitates access to online resources and collaboration tools.

It's important to note that successful integration of technology requires careful planning, professional development for teachers, and addressing any potential access or equity issues. The goal is to leverage technology as a tool to enhance teaching and learning, fostering creativity, collaboration, critical thinking, and problem-solving skills among students.

Virtual reality (VR) and augmented reality (AR) arc immersive technologies that can be utilized in education to create interactive and engaging learning experiences.

1. **Virtual Reality (VR):** VR technology creates a complete digital environment that immerses users in a virtual world. Users wear a VR headset that places them in a simulated 3D environment, cutting off their real-world surroundings. This immersive experience can transport students to different times, places, or scenarios, offering a unique way to learn and explore.

In education, VR can:

- Enhance understanding: Students can explore complex subjects, such as anatomy or physics, by visualizing and interacting with virtual models that would otherwise be difficult to access or visualize. For example, students can explore the human body in 3D or take a virtual tour of historical sites.

- Simulate real-life situations: VR can replicate real-world scenarios, allowing students to practice skills in a safe and controlled environment. For instance, aspiring surgeons can perform virtual surgeries, or engineering students can simulate experiments.

- Foster empathy and perspective-taking: VR experiences can help students develop empathy by placing them in the shoes of others. They can experience different cultures, historical events, or perspectives, promoting understanding and tolerance.

- Gamify learning: VR applications can gamify learning by incorporating game-like elements, challenges, and rewards. This makes the learning process more engaging and enjoyable for students.

2. Augmented Reality (AR): AR technology overlays digital content onto the real-world environment, enhancing it with additional information or virtual objects. Unlike VR, users do not get fully immersed in a virtual world; instead, they experience the physical world with digital augmentations through a device like a smartphone or tablet.

In education, AR can:

- Bring textbooks to life: AR can animate static images in textbooks, transforming them into interactive and dynamic content. Students can scan images with an AR device or app to access additional information, videos, or 3D models related to the topic.

- Create interactive learning experiences: AR enables students to interact with virtual objects or simulations superimposed on the real world. For example, students can conduct virtual chemistry experiments on their desks or explore the solar system by pointing their devices at the night sky.

- Facilitate collaborative learning: AR can promote collaborative learning by allowing multiple students to interact with the same virtual objects or simulations simultaneously. This encourages teamwork, communication, and problem-solving skills.

- Support practical skills development: AR can provide real-time guidance and feedback for hands-on tasks. For instance, AR can overlay instructions and virtual markers on a physical object, helping students assemble or repair things.

VR and AR technologies have the potential to transform the learning experience by making it more immersive, interactive, and engaging, improving students' understanding, retention, and motivation.

Augmented Reality (AR) can greatly enhance hands-on learning experiences by merging the physical world with digital overlays, providing students with interactive and immersive educational opportunities. Here are some ways AR enhances hands-on learning:

1. Visualizing complex concepts: AR can bring abstract or complex concepts to life by overlaying virtual models, animations, or data onto real-world objects. For example, students studying anatomy can use AR to see a 3D representation of the human body and interact with various organs, making the learning experience more tangible and understandable.

2. Simulations and experiments: AR allows students to conduct virtual simulations and experiments in a safe and controlled environment. They can explore physics principles, chemical reactions, or historical events without the need for physical equipment or materials. AR simulations can provide real-time feedback and generate realistic outcomes, promoting critical thinking and problem-solving skills.

3. Enhancing field trips: AR can enhance field trips by providing additional information and interactive experiences. Students can use AR apps on their devices to scan landmarks, historical sites, or natural formations and receive relevant information, virtual tours, or interactive quizzes. This enables a more interactive and engaging exploration of the real world.

4. Augmented textbooks: AR can transform traditional textbooks into interactive learning tools. By scanning specific pages or images, students can access additional multimedia content, such as videos, 3D models, or interactive quizzes, enhancing their understanding and engagement with the material.

5. Virtual overlays and annotations: AR allows students to overlay virtual annotations, labels, or markers onto real-world objects, supporting hands-on experiences. For example, students learning woodworking can use AR

to project cutting guides onto actual wood pieces, improving precision and understanding of techniques.

6. Collaboration and communication: AR can facilitate collaboration and communication among students in hands-on learning scenarios. By sharing AR experiences or collaborating on augmented projects, students can work together, exchange ideas, and learn from each other in an interactive and engaging way.

7. Personalized learning: AR can adapt to students' individual needs and learning styles. By offering interactive and customizable learning experiences, AR allows students to explore topics at their own pace, focus on specific areas of interest, and receive personalized feedback and guidance.

AR technology has the potential to transform traditional hands-on learning, making it more interactive, engaging, and accessible. By seamlessly blending the physical and digital worlds, AR creates unique educational experiences that foster curiosity, critical thinking, and creativity among students.

B. Embracing lifelong learning and adaptability

AR can play a significant role in embracing lifelong learning and adaptability by offering continuous access to educational resources and providing interactive learning experiences for learners of all ages. Here's how AR enhances lifelong learning:

1. Accessible learning anytime, anywhere: AR enables learners to access educational content whenever and wherever they want. Through AR-enabled apps or devices, learners can engage with immersive learning experiences without being limited to a physical classroom or specific time schedules. This flexibility encourages a culture of lifelong learning, as individuals can pursue knowledge and skills at their own pace and convenience.

2. Professional development and upskilling: AR can be utilized to provide interactive training and simulations for professional development and upskilling purposes. Employees can access AR modules to learn new procedures, practice complex tasks, or receive virtual coaching and

feedback. This empowers individuals to adapt to changing technologies and work requirements throughout their careers.

3. Gamification of learning: AR can gamify the learning process, making it more engaging and enjoyable for learners. By incorporating elements like rewards, challenges, and interactive quizzes into AR experiences, lifelong learners are motivated to continue acquiring new knowledge and skills. This gamification approach enhances retention and encourages a positive attitude towards ongoing learning.

4. Virtual mentors and guides: With AR, learners can benefit from virtual mentors and guides who can assist them in their lifelong learning journey. AR applications can provide personalized guidance, explanations, and feedback, allowing learners to receive support and adapt their learning paths based on their specific needs and interests.

5. Interactive language learning: AR can enhance language learning by providing interactive and immersive experiences. Learners can use AR apps to scan objects or environments and instantly receive translations, pronunciations, or contextual information. This makes language acquisition more practical and engaging, enabling individuals to continue learning and practicing new languages throughout their lives.

6. Virtual classrooms and collaborative learning: AR can create virtual classrooms and foster collaborative learning environments regardless of geographical limitations. Learners can connect with peers, teachers, and experts from around the world, sharing AR experiences, working on projects together, and exchanging knowledge and ideas. This promotes a lifelong learning mindset, encouraging individuals to embrace new perspectives and continuously expand their understanding.

7. Personalized learning journeys: AR can personalize learning experiences based on individual preferences, progress, and learning styles. By tracking learners' interactions, AR systems can adapt the content and level of difficulty to match their needs, ensuring an optimal and tailored learning experience throughout their lifelong learning journey.

By leveraging AR technology, individuals can embrace lifelong learning, adapt to new challenges, and acquire new skills and knowledge throughout their lives. AR's ability to offer accessible, interactive, and personalized learning experiences empowers individuals to continuously grow, adapt, and succeed in an ever-changing world.

Virtual classrooms offer several benefits for lifelong learning. Here are a few:

1. Flexibility and accessibility: Virtual classrooms eliminate the constraints of time and physical location, allowing learners to access educational content from anywhere and at any time. This flexibility enables individuals to fit learning into their busy schedules, making lifelong learning more achievable and convenient.

2. Global connectivity: Virtual classrooms connect learners from different parts of the world, creating a diverse and inclusive learning environment. Through online discussions, virtual collaborations, and shared experiences, individuals can broaden their perspectives, learn from different cultures, and engage with a global community of learners.

3. Increased engagement: Virtual classrooms leverage various interactive technologies, including AR, multimedia presentations, and online collaboration tools, to create engaging and dynamic learning experiences. These interactive elements capture learners' attention, enhance their motivation, and encourage active participation, leading to higher levels of engagement and knowledge retention.

4. Personalized learning: Virtual classrooms can be tailored to individual learners' needs and preferences. With online platforms, learners can receive personalized feedback, access additional resources, and progress at their own pace. This customization promotes a more effective and efficient learning experience, allowing learners to focus on areas of interest or areas that require more attention.

5. Enhanced collaboration: Virtual classrooms enable learners to collaborate on group projects, participate in online discussions, and share ideas and insights. Through real-time communication tools and collaborative platforms, learners can exchange knowledge, engage in peer-to-peer learning, and develop teamwork and communication skills. This collaborative aspect of virtual classrooms reinforces lifelong learning by fostering social connections and supporting a sense of community among learners.

6. Continuous learning opportunities: Virtual classrooms provide access to a wide range of educational resources, including recorded lectures, e-books, interactive modules, and expert guest speakers. Learners can revisit materials, engage in self-paced learning, and explore new topics of interest outside the structured curriculum. This continuous access to learning

opportunities promotes lifelong learning and encourages individuals to expand their knowledge and skills beyond traditional boundaries.

7. Cost-effectiveness: Virtual classrooms can offer cost-effective learning options compared to traditional classroom-based education. Learners can save on commuting expenses, accommodation, and other related costs. Additionally, virtual classrooms reduce the need for physical infrastructure and resources, making education more accessible and affordable for a wider range of individuals.

Overall, virtual classrooms provide a flexible, engaging, and inclusive learning environment that supports lifelong learning. The benefits of virtual classrooms empower individuals to pursue education throughout their lives and adapt to new challenges and opportunities.

C. Nurturing creativity, critical thinking, and empathy

In this chapter, we delve into the essential role of education in nurturing creativity, critical thinking, and empathy as we strive towards a future of cooperation and unity. As our society becomes increasingly integrated with advanced AI technologies, it becomes paramount to equip individuals with the skills and mindset necessary to navigate this rapidly changing landscape.

1. Cultivating Creativity: In a world where automation and AI take over routine tasks, creativity becomes a distinguishing human trait. Education should foster environments that encourage creativity, imagination, and originality. By providing opportunities for divergent thinking, open-ended problem solving, and artistic expression, we can unlock and nurture the creative potential within each individual. Emphasizing the importance of creative thinking helps individuals develop innovative solutions to complex challenges, driving progress and adaptation.

2. Fostering Critical Thinking: In an era of information overload, critical thinking skills are imperative. Education should focus on cultivating individuals' ability to analyze and evaluate information critically, question assumptions, and consider different perspectives. By teaching logical reasoning, evidence-based decision making, and the ability to discern reliable sources of information, we empower individuals to make informed choices, contribute to society, and navigate the complexities of an AI-integrated world.

3. Promoting Empathy: As we move towards a more cooperative and interconnected future, empathy emerges as a cornerstone of human interaction. Education plays a pivotal role in developing empathetic individuals who are capable of understanding and valuing the experiences, feelings, and perspectives of others. By incorporating emotional intelligence education, encouraging perspective-taking exercises, and fostering inclusive learning environments, we can foster a culture of compassion, collaboration, and understanding.

4. Interdisciplinary Learning: Integration of various disciplines in education helps students see the interconnectedness of knowledge and foster a holistic understanding of the world. By promoting interdisciplinary learning, individuals gain a broader perspective, enabling them to tackle complex problems from multiple angles and develop innovative solutions. Interdisciplinary education encourages collaboration, nurtures curiosity, and breaks down barriers between traditional subjects, fostering a more comprehensive and interdisciplinary approach to problem-solving.

5. Ethical Considerations: Education also has a crucial role in nurturing ethical awareness and responsible use of technology. As AI plays an increasingly significant role in our lives, it is important to instill ethical values and principles early on. Education should encourage discussions on the ethical implications of AI and guide individuals in making decisions that prioritize human well-being, fairness, and sustainability. By equipping learners with a strong ethical foundation, we can ensure the responsible integration of AI in society.

In this chapter, we explore the vital task of education in nurturing creativity, critical thinking, and empathy. By cultivating these essential skills, we empower individuals to thrive in an AI-integrated world characterized by cooperation and unity. Education becomes the catalyst for personal growth, societal progress, and the advancement of the collective human species.

Critical thinking is of paramount importance in education as it equips individuals with the skills necessary to navigate an increasingly complex and information-driven world. Here are some key reasons why critical thinking holds great significance in education:

1. Active Learning: Critical thinking moves education beyond passive absorption of information and encourages active engagement with the material. It prompts individuals to analyze, evaluate, and question ideas,

fostering a deeper understanding of the subject matter. Instead of accepting information at face value, students learn to examine evidence, consider different perspectives, and draw well-reasoned conclusions.

2. Problem-Solving: Critical thinking is essential for problem-solving. It enables students to approach challenges with a systematic and logical mindset. By breaking down problems, evaluating possible solutions, and anticipating potential consequences, individuals can make informed decisions and develop effective strategies to tackle complex issues. Critical thinking encourages students to consider multiple solutions, think outside the box, and adapt their approaches when necessary.

3. Independent and Informed Thinking: In a world overflowing with information, critical thinking empowers individuals to be discerning and independent thinkers. It encourages students to question assumptions, evaluate biases, and critically analyze the credibility and validity of sources. By honing the ability to differentiate between reliable information and misinformation, individuals can make better-informed decisions and form their own opinions based on evidence and rational analysis.

4. Effective Communication: Critical thinking is closely linked to effective communication skills. It enables individuals to articulate their thoughts clearly and persuasively, supporting their claims with logical reasoning and evidence. By developing critical thinking skills, students improve their ability to convey complex ideas, engage in respectful debates, and collaborate effectively with others. Critical thinking also fosters active listening skills, enabling individuals to understand and evaluate different perspectives.

5. Lifelong Learning: Critical thinking is not limited to specific subjects or disciplines; it is a transferable skill that is valuable in all areas of life. Education that emphasizes critical thinking equips individuals with the tools to continue learning and adapting in an ever-evolving world. Critical thinkers are more likely to seek out new knowledge, question their own assumptions, and stay open to different viewpoints, enabling continuous personal and intellectual growth.

In summary, critical thinking plays a vital role in education by fostering active learning, problem-solving abilities, independent thinking, effective communication, and a lifelong pursuit of knowledge. By honing critical thinking skills, education empowers individuals to navigate a vast amount of information, make informed decisions, and contribute meaningfully to society.

Integrating critical thinking into different subjects can greatly enhance students' learning experiences and their overall development. Here are some examples of how critical thinking can be incorporated into various subjects:

1. Language Arts/English: Encourage students to critically analyze literary works by asking them to identify themes, evaluate characters' motivations, and analyze the author's use of language and literary devices. Have them express their opinions about the text, supporting their arguments with evidence from the text.

2. Science: Promote critical thinking in science by asking students to design experiments, analyze data, and draw conclusions based on evidence. Encourage them to question scientific theories and explore alternative explanations. Have them evaluate the reliability of sources and critically assess research studies.

3. Mathematics: In math, encourage students to solve complex problems using critical thinking skills. Instead of simply memorizing formulas, have them understand the underlying concepts and apply logical reasoning to arrive at solutions. Encourage them to explore multiple approaches and justify their answers.

4. Social Studies/History: Promote critical thinking in social studies by encouraging students to analyze primary and secondary sources. Have them evaluate bias and reliability of sources, research different perspectives on historical events, and construct evidence-based arguments. Encourage discussions that challenge preconceived notions and explore cause-and-effect relationships.

5. Arts and Humanities: Incorporate critical thinking in arts and humanities by having students critically evaluate and interpret works of art, music, or literature. Encourage them to express their interpretations and support them with evidence. Foster discussions that explore the cultural and historical contexts of artistic creations.

6. Problem-Based Learning: Integrate critical thinking across subjects through problem-based learning. Present students with real-world problems or scenarios that require them to employ critical thinking skills from various disciplines to find solutions. This approach fosters interdisciplinary thinking, collaboration, and creativity.

7. Classroom Discussions and Debates: Engage students in meaningful discussions and debates within all subject areas. Encourage them to ask questions, justify their opinions with evidence, listen actively to different viewpoints, and respectfully challenge ideas. Create an environment that values critical thinking and fosters intellectual curiosity.

Remember, it's important to provide guidance and scaffold students' critical thinking skills gradually, ensuring they understand the process and have the necessary tools to analyze, evaluate, and think critically. By integrating critical thinking into different subjects, educators can help students develop essential skills that will benefit them throughout their academic journey and beyond.

To ensure that intelligence enhancement technologies do not reinforce biases or discriminatory practices, several measures can be taken:

1. Ethical guidelines and standards: Develop clear and comprehensive ethical guidelines that explicitly state the importance of avoiding biases and discriminatory practices in the design, development, and deployment of intelligence enhancement technologies. These guidelines should be prioritized and followed by developers, researchers, and policymakers involved in the field.

2. Diverse and inclusive development teams: Foster diverse and inclusive development teams that consist of individuals from different backgrounds, experiences, and perspectives. This allows for a more comprehensive understanding of potential biases and helps in designing technologies that are more inclusive and equitable.

3. Robust data collection: Ensure that data collection processes are unbiased and encompass diverse populations. Collecting representative and inclusive datasets helps reduce the risk of biased algorithms or models being used in intelligence enhancement technologies. Avoid relying on data that may reinforce existing biases or discriminatory practices.

4. Transparent algorithms and decision-making: Make the algorithms and decision-making processes used in intelligence enhancement technologies transparent and understandable. This transparency allows for scrutiny and identification of any biases or discriminatory factors that may be present in the technology. By making the decision-making process transparent, it becomes easier to detect and correct any potential biases.

5. Continuous monitoring and evaluation: Conduct regular and ongoing monitoring and evaluation of intelligence enhancement technologies to identify any biases that may emerge over time. This monitoring should involve diverse stakeholders, including ethicists, researchers, user communities, and affected populations. If biases or discriminatory practices are detected, they should be addressed promptly and transparently.

6. User feedback and engagement: Involve users and affected populations in the design and development process of intelligence enhancement technologies. Gathering feedback from diverse users helps identify potential biases, discriminatory practices, or unintended consequences that may arise from using the technology. User engagement also ensures that the technology is inclusive, user-friendly, and aligned with the needs and values of diverse individuals and communities.

7. Regulatory frameworks: Establish regulatory frameworks that explicitly address the avoidance of biases and discriminatory practices in intelligence enhancement technologies. These frameworks should include guidelines for accountability, transparency, and safeguards against biases or discriminatory effects. Independent oversight and auditing mechanisms can help ensure compliance with these regulations.

By implementing these approaches, intelligence enhancement technologies can strive to minimize biases and discriminatory practices, promoting fairness, inclusivity, and equity. It requires a conscious effort to develop technologies that uphold ethical standards, involve diverse perspectives, and prioritize the well-being and rights of all individuals.

Diverse development teams are of paramount importance in the field of intelligence enhancement technologies for several reasons:

1. Inclusive perspectives: Diverse development teams bring together individuals with different backgrounds, experiences, and perspectives. This diversity of viewpoints helps in considering a wide range of needs, preferences, and potential impacts of intelligence enhancement technologies on various user groups. It ensures that the technology is designed with inclusivity in mind and addresses the requirements of a diverse user base.

2. Minimizing biases: Biases can inadvertently be introduced into technology design, development, and decision-making processes. By having a diverse team, biases can be identified and addressed more effectively.

Different team members may be aware of their own biases or recognize biases that others might overlook, leading to a more balanced approach in the development of intelligence enhancement technologies.

3. Improved problem-solving and innovation: When people from different backgrounds and perspectives come together, they bring unique knowledge, skills, and insights. This diversity fosters creative problem-solving and stimulates innovation. Diverse teams are more likely to approach challenges from different angles and generate a wider range of ideas, ultimately leading to more effective and impactful intelligence enhancement technologies.

4. User-centered design: The involvement of diverse development teams helps ensure that intelligence enhancement technologies are relevant and user-centered. Different team members can provide valuable input based on their lived experiences, enabling the technology to better cater to the needs, preferences, and values of diverse user groups. This user-centered approach enhances the usability and effectiveness of the technology.

5. Avoiding unintended consequences: Intelligence enhancement technologies can have unintended consequences, such as exacerbating inequalities or reinforcing stereotypes. By having a diverse team, potential risks and unintended consequences can be identified more effectively. Team members with different perspectives can provide valuable insights into potential issues and help mitigate unintended negative impacts.

6. Representation and fairness: Diverse development teams promote fair representation in the development process. They ensure that the voices and perspectives of marginalized or underrepresented groups are considered and incorporated. This helps prevent the creation of technologies that inadvertently perpetuate existing inequalities or discriminatory practices.

7. Enhanced cultural competency: Intelligence enhancement technologies are used in various cultural contexts. A diverse team can provide cultural competency and sensitivity, considering the social and cultural implications of the technology across different communities. This helps ensure that the technology respects and aligns with diverse cultural values and norms.

In summary, diverse development teams are essential for developing inclusive, unbiased, and effective intelligence enhancement technologies. They contribute to a more comprehensive understanding of user needs, foster innovation, mitigate biases, and minimize unintended negative consequences. Embracing diversity in the development process ultimately leads to technologies that have a broader positive impact on society as a whole.

Chapter 16: Building an Inclusive Society

A. Ensuring inclusivity in the age of technological advancements

With the rapid advancement of technology, it is imperative to ensure that inclusivity remains at the forefront of societal development. Technological advancements have the potential to either foster inclusivity or exacerbate existing inequalities. In this chapter, we will explore how to ensure inclusivity in the age of technological advancements and promote equal access and participation for all.

1. Digital Accessibility: As technology becomes increasingly integrated into various aspects of our lives, it is crucial to prioritize digital accessibility. This involves designing and developing digital platforms, websites, and applications that are accessible to individuals with different disabilities. Implementing features like screen reader compatibility, alternative text for images, closed captioning, and keyboard navigation can greatly enhance accessibility and ensure that everyone can benefit from these technological advancements.

2. Bridging the Digital Divide: While technology has the potential to provide opportunities for all, it is essential to address the digital divide that exists between different communities. Many individuals, particularly in marginalized communities, may not have access to internet connectivity or the necessary devices. Governments, organizations, and educational institutions should work together to provide affordable or free internet access and devices to those who lack them. This will bridge the digital divide and ensure that everyone has equal access to information, communication, and opportunities.

3. Digital Literacy: Promoting digital literacy among all individuals is key to ensuring inclusivity in the age of technological advancements. This involves providing training and education to individuals who may be unfamiliar with technology or lack the necessary skills to utilize it effectively. By empowering individuals with digital literacy skills, we can enable them to fully participate in the digital world and take advantage of the opportunities it offers.

4. **Addressing Bias and Discrimination:** Technological advancements can inadvertently perpetuate bias and discrimination if not carefully scrutinized. Algorithms and artificial intelligence systems can inherit the biases present in the data used to train them, leading to discriminatory outcomes. It is essential to develop and enforce ethical guidelines for the use of technology, ensuring that systems are designed and implemented in a way that promotes fairness and inclusivity. Regular audits and evaluations can help identify and address potential biases within technological systems.

5. **Collaboration and Engagement:** Building an inclusive society in the age of technological advancements requires collaboration and engagement from all stakeholders. Governments, private organizations, educational institutions, and individuals must work together to ensure that everyone has a voice in shaping the direction of technology. Engaging diverse perspectives in the development, implementation, and evaluation of technological advancements can help prevent exclusion and ensure that the needs of all individuals are considered.

By prioritizing digital accessibility, bridging the digital divide, promoting digital literacy, addressing bias and discrimination, and fostering collaboration and engagement, we can build an inclusive society in the age of technological advancements. It is our collective responsibility to harness the potential of technology to create a world in which everyone can fully participate and benefit, regardless of their background or abilities.

The role of government in promoting digital inclusivity is crucial in ensuring that technology benefits all members of society. Here are some ways in which governments can play a significant role:

1. **Policy and Regulation:** Governments can establish policies and regulations that promote digital inclusivity. They can enforce laws that require public facilities, transportation systems, and government websites to be accessible to individuals with disabilities. They can also regulate internet service providers to ensure affordable and reliable internet access for all communities, especially those in underserved areas.

2. **Funding and Infrastructure:** Governments can allocate funding and resources to bridge the digital divide. This can involve investing in the development of broadband infrastructure in remote or underprivileged areas, providing subsidies or grants for low-income individuals to access internet services, and supporting initiatives that promote technology adoption and digital literacy in marginalized communities.

3. Digital Skills Training: Governments can collaborate with educational institutions, nonprofits, and private organizations to offer digital skills training programs. These programs could include basic computer literacy, online safety, and specific digital skills needed for employment or entrepreneurship. By equipping individuals with the necessary skills, governments can empower them to participate fully in the digital economy.

4. Accessible Public Services: Governments should ensure that their own digital platforms and services are accessible to all. This includes websites, mobile apps, and online portals for accessing public services such as healthcare, education, social welfare, and government applications. Implementing universal design principles and conducting regular accessibility audits can ensure that government services are inclusive and accessible.

5. Collaboration with Stakeholders: Governments can collaborate with stakeholders from various sectors, including technology companies, civil society organizations, and advocacy groups. By working together, they can identify barriers to digital inclusivity, develop solutions, and promote best practices. Governments can also engage in dialogue with marginalized communities and involve them in decision-making processes related to technology and digital access.

6. Monitoring and Evaluation: Governments should monitor and evaluate the progress of their digital inclusivity initiatives to ensure effectiveness and identify areas for improvement. This can involve collecting data on internet access, digital skills, and usage patterns to identify disparities and implement targeted interventions. Regular evaluation can help governments refine their strategies and policies to better address the needs of underserved populations.

Governments have a critical role to play in creating an environment where digital inclusivity is prioritized and accessibility barriers are minimized. By implementing policies, providing infrastructure and resources, promoting digital skills training, ensuring accessible public services, collaborating with stakeholders, and monitoring progress, governments can contribute significantly to building an inclusive society in the digital age.

Collaboration between government and technology companies is important for driving innovation, addressing societal challenges, and promoting the overall well-being of citizens. Here are some aspects of collaboration between these entities:

1. **Policy Development:** Governments rely on the expertise of technology companies to develop effective policies and regulations. Technology companies possess valuable insights into emerging technologies, industry best practices, and market trends. Collaborating with these companies allows governments to create policies that strike a balance between promoting innovation and protecting the interests of society.

2. **Infrastructure Development:** Technology companies often have the resources, technical expertise, and infrastructure required to deploy and maintain digital infrastructure. Governments can collaborate with technology companies to build and upgrade broadband networks, set up data centers, and develop smart city initiatives. This collaboration helps ensure that citizens have access to reliable and high-speed internet services, as well as advanced digital tools and solutions.

3. **Public-Private Partnerships:** Governments can form public-private partnerships (PPPs) with technology companies to jointly tackle complex societal issues. PPPs leverage the strengths of both sectors, combining government resources and regulatory authority with the industry's technical expertise and efficiency. This collaboration can address challenges such as healthcare technology, cybersecurity, digital inclusion, transportation, and sustainable development.

4. **Research and Development:** Governments often partner with technology companies to foster research and development (R&D) initiatives. Through collaborative R&D projects, new solutions and technologies can be developed to address key societal challenges. Funding programs, grants, and tax incentives are often provided to encourage joint R&D efforts, leading to technological advancements and economic growth.

5. **Data Sharing and Analytics:** Collaboration between governments and technology companies allows for data sharing and analytics capabilities. Governments can access valuable data insights from technology companies to inform evidence-based policymaking and decision-making processes. This collaboration can also help identify trends, patterns, and potential solutions for various issues, such as urban planning, public health, and transportation.

6. **Ethical and Responsible Technology Development:** Governments and technology companies can collaborate to establish ethical frameworks and guidelines for the responsible development and use of technology. This includes ensuring data privacy, cybersecurity, and fairness in algorithms and artificial intelligence systems. By working together, they can create

policies and standards that protect users' rights and maintain public trust in technology.

Collaboration between government and technology companies is a symbiotic relationship that promotes innovation, addresses societal challenges, and enhances the well-being of citizens. By leveraging each other's strengths and expertise, these entities can create a conducive environment for technological advancements and societal progress.

Collaboration between governments and technology companies offers a multitude of benefits for both parties and society as a whole. Here are some key advantages:

1. Innovation and Technological Advancement: By collaborating, governments and technology companies can drive innovation and technological advancement. Governments often lack the resources and expertise of tech companies, and partnering with them allows access to cutting-edge technologies, research and development, and industry knowledge. This helps accelerate the development and deployment of new solutions that can benefit society.

2. Efficient Service Delivery: Collaboration enables governments to leverage the efficiency and expertise of technology companies to improve service delivery. By incorporating digital tools, automation, and data analytics, governments can streamline their operations, enhance public services, and deliver better outcomes to citizens. This can range from online portals for government services to digital platforms for citizen engagement or e-governance initiatives.

3. Economic Growth and Job Creation: Government-technology collaboration can stimulate economic growth and create employment opportunities. When governments support and partner with technology companies, it leads to the expansion of the tech industry. This results in a boost to the local economy, increased investments, and the creation of new jobs in technology-related fields.

4. Addressing Societal Challenges: Collaborative efforts between governments and technology companies can help address societal challenges and improve the well-being of citizens. By combining resources, expertise, and data-driven insights, they can develop innovative solutions for healthcare, education, transportation, sustainability, public safety, and more. This collaboration allows for a comprehensive and effective approach to tackling complex problems.

5. Policy Development and Regulation: Governments rely on technology companies' expertise to develop informed policies and regulations for emerging technologies. Through collaboration, governments gain a better understanding of technological advancements and the potential impacts on society. This can lead to well-balanced policies that foster innovation while considering privacy, security, ethics, and other societal concerns.

6. Data-Driven Decision Making: Collaboration enables the exchange of data and knowledge between governments and technology companies. Governments can leverage the data insights and analytics capabilities of tech companies to make informed decisions. This helps policymakers identify trends, anticipate challenges, and develop evidence-based policies that have a positive impact on citizens.

7. Infrastructure Development: Governments often collaborate with technology companies to develop and upgrade critical digital infrastructure such as broadband networks, smart cities, and data centers. These collaborations help improve connectivity, digital inclusion, and overall infrastructure quality, enabling citizens to access the benefits of technology more easily.

Collaboration between governments and technology companies holds great potential for driving progress, improving public services, and addressing societal challenges in an innovative and efficient manner. By combining their resources, expertise, and insights, both parties can achieve better outcomes for society as a whole.

While collaboration between governments and technology companies offers many benefits, there are also challenges that need to be navigated. Here are some of the common challenges faced in government-technology collaboration:

1. Varying Goals and Priorities: Governments and technology companies often have different goals and priorities. Governments focus on public service delivery, policymaking, and regulatory compliance, while technology companies prioritize innovation, profit, and market competition. Aligning these diverse objectives can be challenging and requires effective communication and negotiation.

2. Bureaucracy and Decision-Making Processes: Government agencies often have complex bureaucratic processes and lengthy decision-making cycles. On the other hand, technology companies are accustomed to more

agile and fast-paced environments. The disparity in decision-making processes can slow down collaboration efforts, hinder innovation, and cause frustration for both parties.

3. Privacy and Security Concerns: Collaboration between governments and technology companies involves the sharing and analysis of vast amounts of data, raising privacy and security concerns. Protecting sensitive information and ensuring data privacy while harnessing the power of analytics requires robust security measures and clear agreements on data sharing and usage. Balancing the need for data-driven insights with privacy protection can be a significant challenge.

4. Procurement and Contracting Challenges: The process of procuring technology solutions and entering into contracts with technology companies can be complex and time-consuming for governments. Ensuring transparency, fairness, and cost-effectiveness in procurement processes, while also meeting the specific needs of government agencies, can be a challenge. Delays in procurement and delivery can impact project timelines and outcomes.

5. Technological Capacity and Skill Gap: Governments may face challenges in building and maintaining technological capacity and expertise. Keeping up with rapidly evolving technologies and integrating them into existing government systems can be daunting. Additionally, the skill gap between government employees and technology professionals can hinder effective collaboration and implementation of technology initiatives.

6. Regulatory Hurdles and Legal Frameworks: The regulatory landscape for technology varies across jurisdictions, which can create challenges in government-technology collaboration. Governments must navigate legal frameworks related to data protection, intellectual property, cybersecurity, and privacy, among others. Harmonizing regulations and ensuring compliance can be complex, particularly when collaborating across borders.

7. Resistance to Change and Organizational Culture: Governments are often characterized by hierarchical structures and entrenched work practices. Introducing new technologies and ways of working may face resistance from within government agencies. Overcoming resistance to change and fostering a culture of innovation can be challenging but is crucial for successful collaboration.

Despite these challenges, government-technology collaboration can yield substantial benefits. Recognizing and actively addressing these challenges through effective communication, clear agreements, capacity building, and collaboration frameworks can help ensure successful partnerships between governments and technology companies.

B. Addressing socio-economic disparities through cooperation

Addressing socio-economic disparities requires collaboration between governments, organizations, and communities. Here are some ways cooperation can be effective in tackling these disparities:

1. Policy Development and Implementation: Governments can work with organizations and communities to develop and implement policies that address socio-economic disparities. This can include initiatives to reduce income inequality, improve access to education and healthcare, provide affordable housing, and enhance social safety nets. Collaboration can ensure that policies are holistic, responsive, and tailored to specific socio-economic challenges.

2. Resource Allocation and Investment: Collaborative efforts can focus on effective resource allocation and investment in disadvantaged communities. Governments can partner with organizations and businesses to provide financial support, infrastructure development, and access to capital for small businesses. By working together, they can leverage resources and expertise to create sustainable economic opportunities and uplift marginalized communities.

3. Education and Skill Development: Cooperation between governments, educational institutions, and employers is essential to address educational and skill gaps. By aligning curricula with industry needs, providing vocational training, and offering internships or apprenticeships, collaborative efforts can equip individuals with the necessary skills for employment and economic advancement.

4. Technology and Digital Inclusion: Cooperation can promote digital inclusion to bridge the digital divide. This involves ensuring affordable internet access, providing digital literacy programs, and making technology more accessible to underserved communities. Governments, technology companies, and non-profit organizations can work together to

develop initiatives that improve digital skills and facilitate digital inclusion for all.

5. Community Engagement and Empowerment: Collaborative efforts should involve meaningful community engagement and empowerment. Governments and organizations can partner with community leaders, grassroots organizations, and local residents to understand the specific needs and challenges faced by communities. Through active participation, communities can take ownership of their development, advocate for their rights, and contribute to decision-making processes.

6. Data-driven approaches: Cooperation can facilitate data sharing and analysis to better understand socio-economic disparities and develop evidence-based solutions. Governments, researchers, and organizations can collaborate to collect, analyze, and share data on various socio-economic indicators. This collaborative effort can help identify trends, disparities, and effective strategies to address the root causes of socio-economic inequalities.

7. Multisector Partnerships: Collaboration across sectors is essential for sustainable and comprehensive approaches to socio-economic disparities. Governments, non-profit organizations, businesses, and academic institutions can come together to develop joint initiatives, share resources, and pool expertise. These partnerships can foster innovation, leverage diverse perspectives, and create a collective impact that goes beyond the capabilities of individual actors.

By embracing cooperation and collaboration, governments, organizations, and communities can work together to address socio-economic disparities more effectively. Through shared goals, coordinated actions, and inclusive decision-making, we can create a more equitable and inclusive society.

Expanding access to quality education for marginalized groups is crucial for reducing socio-economic disparities. Here are some strategies to achieve this goal:

1. Eliminating Barriers: Identify and eliminate barriers that prevent marginalized groups from accessing education. This could involve addressing issues such as discrimination, socio-economic barriers, lack of infrastructure, and cultural biases. Governments and organizations can work together to ensure that education is inclusive and accessible to all, regardless of race, ethnicity, gender, disability, or socio-economic background.

2. Affirmative Action and Scholarships: Implement affirmative action policies and scholarships specifically designed to increase the representation of marginalized groups in educational institutions. These initiatives can help level the playing field by providing opportunities for those who face historical disadvantages. Scholarships can support economically disadvantaged students, while affirmative action can ensure fair representation in admissions and hiring.

3. Community Engagement: Engage with marginalized communities to understand their specific educational needs and challenges. Collaborate with community leaders, parents, and local organizations to develop tailored solutions. Ensure that education programs are culturally sensitive and take local context into account.

4. Early Childhood Education: Encourage early childhood education programs that are accessible and of high quality. By investing in early education, we can provide a strong foundation for marginalized children, helping them develop the necessary skills and competencies needed for success later in life.

5. Teacher Training and Support: Provide training and support for teachers working in marginalized areas to ensure they have the necessary skills and knowledge to effectively teach diverse student populations. Training programs should focus on cultural competency, inclusion, and understanding the unique needs of marginalized students.

6. Addressing Socio-Economic Factors: Recognize and address the socio-economic factors that hinder educational attainment for marginalized groups. This could involve providing financial assistance, school feeding programs, transportation support, and additional resources to help students overcome barriers related to poverty.

7. Technology and Distance Learning: Leverage technology to expand access to education, particularly in remote areas or areas with limited resources. Invest in infrastructure, provide access to computers and the internet, and promote distance learning initiatives. This can help bridge the education gap and ensure marginalized groups have equal opportunities to learn.

8. Partnerships and Collaboration: Foster partnerships between governments, education institutions, private sector organizations, and non-profit groups to pool resources, share best practices, and create innovative

solutions. Collaborative efforts can help address the complex challenges associated with expanding access to quality education for marginalized groups.

By prioritizing and implementing these strategies, we can work towards creating a more equitable education system that provides equal opportunities for all, regardless of their socio-economic background or other marginalized identities.

Supporting non-traditional education pathways is essential for accommodating diverse learning styles and expanding access to education beyond the traditional classroom setting. Here are some ways to support non-traditional education pathways:

1. Recognition and Validation: Recognize and validate non-traditional forms of education, such as online courses, vocational training, apprenticeships, and self-directed learning. This could involve creating systems for accrediting and certifying skills obtained outside of traditional educational institutions to ensure they are valued and accepted by employers and academic institutions.

2. Flexible Learning Options: Provide flexible learning options that allow individuals to tailor their education to their specific needs and circumstances. This could include part-time and evening classes, weekend programs, online courses, and blended learning models that combine online and in-person instruction.

3. Mentorship and Guidance: Offer mentorship and guidance programs to support individuals pursuing non-traditional education pathways. Mentors can provide valuable advice, encouragement, and industry-specific insights, helping learners navigate their chosen paths and connect with opportunities.

4. Financial Support: Offer financial assistance and scholarships specifically aimed at supporting individuals pursuing non-traditional education pathways. This can help mitigate financial barriers and make education more accessible to a wider range of learners.

5. Collaboration with Industry: Foster partnerships between educational institutions and industries to develop non-traditional educational programs that align with industry needs. By involving industry experts in curriculum development and providing work-integrated learning opportunities,

learners can gain practical skills and experiences that enhance their employability.

6. Access to Resources: Ensure individuals have access to educational resources and materials necessary for non-traditional learning pathways. This could involve providing online libraries, open educational resources, and affordable or free access to learning platforms and tools.

7. Recognition of Prior Learning: Establish mechanisms for recognizing and valuing prior learning experiences and skills acquired outside of formal education. This can help individuals transition into non-traditional education pathways by leveraging their existing knowledge and expertise.

8. Networking and Community Building: Create platforms and communities that facilitate networking and collaboration among individuals pursuing non-traditional education pathways. This can help learners connect with peers, mentors, and professionals in their chosen fields, fostering a supportive and collaborative learning environment.

Supporting non-traditional education pathways promotes inclusivity, lifelong learning, and the development of skills that are relevant in today's rapidly changing world. By embracing diverse educational options, we can empower individuals to pursue their passions, acquire valuable knowledge and skills, and contribute to society in meaningful ways.

The future possibility of intelligence increase through gene-editing and technological means like Neuralink presents interesting opportunities and challenges. While it has the potential to enhance cognitive abilities and development speed, ensuring equal opportunities and building a better world for everyone still requires careful consideration and efforts. Here are a few points to consider:

1. Ethical considerations: The ethical implications of intelligence enhancement technologies must be carefully examined. It is important to address concerns related to consent, potential misuse, and unintended consequences. Decision-making about such technologies should involve comprehensive ethical discussions and strict regulations to ensure responsible and equitable use.

2. Accessibility and equality: To build a better world, it is essential to ensure that advancements in intelligence enhancement are accessible to all, regardless of socio-economic background, geographic location, or other

factors. Efforts should be made to prevent disparities and create opportunities for everyone to benefit from these advancements.

3. Education and training: As intelligence is enhanced, it will be crucial to provide adequate education and training to help individuals adapt and utilize their enhanced cognitive abilities effectively. This means investing in programs that promote critical thinking, creativity, and social skills alongside technical knowledge.

4. Addressing societal challenges: While increased intelligence may bring significant positive impacts, it is important to address the potential challenges it could create, such as disparities in cognitive abilities between individuals and potential social divisions. We must actively work towards fostering inclusivity and social cohesion in this advancing landscape.

5. Continuous monitoring and evaluation: As intelligence enhancement technologies progress, ongoing monitoring and evaluation are necessary to ensure the well-being, safety, and ethical use of these technologies. Policies must be in place to regularly assess the impacts and regulate their development and deployment.

It is essential to approach intelligence enhancement technologies with caution, emphasizing ethical considerations, inclusivity, and accessibility. By promoting responsible innovation and addressing potential challenges, we can strive towards a future where enhanced intelligence contributes to building a better world for everyone, with equal opportunities for personal and societal growth.

While intelligence enhancement has the potential for various benefits, it is crucial to consider the potential risks that come with it. Here are some of the key risks associated with intelligence enhancement:

1. Unequal access: One of the significant risks is the potential for unequal access to intelligence-enhancing technologies. If these technologies are expensive or limited to certain groups, it could widen existing social and economic inequalities, creating a divide between those who have access to enhanced intelligence and those who do not.

2. Ethical concerns: The ethical implications of intelligence enhancement need careful consideration. Questions may arise regarding the boundaries of enhancement, consent, and the potential unintended consequences that could arise from altering cognitive abilities. These ethical dilemmas need to

be addressed to ensure responsible and ethical use of intelligence enhancement technologies.

3. Psychological well-being: Intelligence enhancement could put individuals under pressure to constantly achieve higher levels of intelligence. This may lead to psychological strain, anxiety, and even mental health issues as people strive to meet the heightened expectations associated with enhanced intelligence.

4. Social and economic disruption: The introduction of intelligence enhancement technologies may disrupt existing social and economic systems. There could be potential job displacement as automation and AI technologies advance alongside intelligence enhancement. This could lead to increased inequality and economic disparities if not managed effectively.

5. Security and privacy concerns: Enhancing cognitive abilities through technologies like brain-computer interfaces raises concerns about security and privacy. As personal thoughts and data become accessible through interconnected networks, there is a need to safeguard individuals' privacy and protect against potential breaches or misuse of this information.

6. Unintended consequences: The long-term effects and unintended consequences of intelligence enhancement are still uncertain. Altering cognitive abilities could have unpredictable outcomes, affecting not just individuals but also social dynamics, relationships, and societal structures. It is essential to thoroughly research and understand the potential long-term impacts before widespread adoption.

7. Dependency on technology: Increased reliance on intelligence enhancement technologies may create dependencies and reduce self-reliance in critical thinking and problem-solving. This could potentially result in diminishing skills without the aid of technology or over-reliance on certain cognitive functions.

To mitigate these risks, it is crucial to have robust regulatory frameworks, ethical guidelines, comprehensive research, and ongoing monitoring of the impact of intelligence enhancement technologies. Responsible development and deployment are necessary to address these risks and ensure that intelligence enhancement promotes overall well-being, equity, and a positive impact on society.

C. Celebrating diversity and leveraging collective wisdom

Celebrating diversity and leveraging collective wisdom are essential principles to consider when exploring intelligence enhancement technologies. Here's why:

1. Diverse perspectives: Celebrating diversity ensures that intelligence enhancement technologies are designed to cater to a wide range of individuals with varying backgrounds, experiences, and cognitive abilities. Embracing diversity fosters inclusivity and prevents the development of technologies that only benefit a narrow group of people.

2. Different forms of intelligence: Intelligence is not limited to a single dimension or parameter. It encompasses various aspects such as emotional intelligence, social intelligence, creativity, and critical thinking. By celebrating diversity, we recognize and appreciate different forms of intelligence, ensuring that intelligence enhancement is not limited to just one aspect but embraces the richness of human capabilities.

3. Cultural and societal contributions: Diversity and cultural richness bring a multitude of perspectives and knowledge systems. By celebrating this diversity, intelligence enhancement can tap into the collective wisdom of different cultures and societies. This inclusivity allows for a more holistic and comprehensive approach to solving complex problems and advancing human understanding.

4. Collaborative problem-solving: Leveraging collective wisdom acknowledges that no single individual has all the answers. By fostering collaboration and teamwork, intelligence enhancement can benefit from the collective intelligence of groups, encouraging the sharing of ideas, perspectives, and insights. This collaborative approach can lead to more effective problem-solving and innovation.

5. Avoiding biases and limitations: Intelligence enhancement technologies should be designed in a way that prevents the reinforcement of biases or discriminatory practices. Celebrating diversity ensures that technologies are developed with a focus on fairness, inclusivity, and avoiding the amplification of existing inequalities or prejudices.

6. Respecting individual choices: Celebrating diversity means respecting individual choices when it comes to intelligence enhancement. Some individuals may choose not to enhance their intelligence, and that choice

should be respected. People should have the autonomy to decide whether they want to pursue intelligence enhancement and to what extent.

By celebrating diversity and leveraging collective wisdom, intelligence enhancement technologies can be harnessed to create a more inclusive, equitable, and thriving society. Embracing diverse perspectives and knowledge systems leads to better solutions, promotes fairness, and respects individual autonomy and choices.

Chapter 17: Environmental Stewardship

A. Seeking sustainable solutions for the planet

Environmental stewardship is crucial in addressing the challenges our planet faces today. Chapter 17 focuses on seeking sustainable solutions to protect and preserve our environment. Here are some key points on environmental stewardship:

1. Conservation and preservation: Environmental stewardship involves conserving and preserving natural resources, biodiversity, and ecosystems. It emphasizes responsible management and protection of the environment to maintain its integrity for future generations.

2. Sustainable development: Stewardship emphasizes the importance of sustainable development, which balances economic growth with environmental protection. It advocates for using resources in a way that meets the needs of the present without compromising the ability of future generations to meet their own needs.

3. Climate change mitigation: As stewards of the environment, it is crucial to address climate change. This involves reducing greenhouse gas emissions, transitioning to renewable energy sources, promoting energy efficiency, and implementing climate change adaptation strategies.

4. Waste reduction and recycling: Environmental stewardship encourages waste reduction and recycling practices. Minimizing waste and promoting recycling help conserve resources, reduce pollution, and decrease the strain on landfills and natural habitats.

5. Conservation of water resources: Stewardship involves responsible water management practices. This includes promoting water conservation,

reducing water pollution, and implementing sustainable water use strategies in agriculture, industry, and households.

6. Biodiversity protection: Stewardship focuses on protecting and conserving biodiversity. It involves efforts to preserve habitats, prevent species extinction, and promote sustainable land-use practices that ensure the long-term survival of diverse ecosystems and the species within them.

7. Education and awareness: Environmental stewardship involves raising awareness and educating individuals about the importance of environmental conservation and sustainability. It encourages people to make informed choices and take actions to protect the environment in their daily lives.

8. Collaboration and advocacy: Stewardship requires collaboration between governments, organizations, communities, and individuals. By working together, advocating for environmentally responsible policies, and implementing sustainable practices, we can make a significant positive impact on the environment.

Environmental stewardship plays a pivotal role in ensuring the long-term health and sustainability of our planet. It is a collective responsibility that requires everyone's commitment to protect and preserve our natural resources for current and future generations. Through sustainable solutions and conscious decisions, we can create a more environmentally sustainable and resilient future.

There are numerous examples of sustainable development practices that aim to balance economic growth with environmental protection. Here are a few key examples:

1. Renewable energy: Transitioning from fossil fuels to renewable energy sources such as solar, wind, hydro, and geothermal power helps reduce greenhouse gas emissions and decreases dependence on finite resources. Additionally, it promotes a more sustainable and cleaner energy future.

2. Energy efficiency: Implementing energy-efficient technologies and practices in buildings, industries, and transportation can significantly decrease energy consumption. This includes using energy-efficient appliances, improving insulation, promoting public transportation, and encouraging energy-conscious behavior.

3. Sustainable agriculture: Practices such as organic farming, agroforestry, and permaculture minimize the use of synthetic fertilizers, pesticides, and genetically modified organisms. They aim to protect soil health, conserve water resources, and promote biodiversity while ensuring sustainable food production.

4. Waste management: Implementing effective waste management practices, including recycling, composting, and waste reduction, helps divert waste from landfills and reduces greenhouse gas emissions. This promotes the efficient use of resources while minimizing pollution and environmental degradation.

5. Green building and urban planning: Incorporating sustainable design principles in architecture and urban planning can lead to energy-efficient buildings, green spaces, and reduced ecological footprints. This includes designing for passive solar heating, using sustainable building materials, and creating walkable and bike-friendly communities.

6. Conservation and restoration of ecosystems: Protecting and restoring natural habitats and ecosystems is essential for biodiversity conservation. This can be achieved through initiatives like reforestation, habitat preservation, and the creation of protected areas. Restoring degraded ecosystems helps enhance their resilience and contributes to carbon sequestration.

7. Water management: Sustainable water management practices include water conservation techniques, rainwater harvesting, and implementing efficient irrigation systems. Additionally, reducing water pollution and promoting responsible water use in industries, agriculture, and households are vital for preserving this finite resource.

8. Sustainable transportation: Encouraging the use of public transportation, promoting electric and hybrid vehicles, and investing in bike lanes and pedestrian-friendly infrastructure help reduce transportation-related emissions and congestion. These practices increase mobility while decreasing the environmental impact of transportation.

These examples highlight the diversity of sustainable development practices that can be implemented across various sectors. By adopting such practices, we can move towards a more sustainable and environmentally friendly future.

Implementing sustainable development practices faces several challenges. Here are some common obstacles that need to be addressed:

1. Economic considerations: Sustainable development practices often require upfront investments and may have higher initial costs compared to traditional, less sustainable alternatives. The economic viability and financial feasibility of sustainable projects can be a challenge, especially for smaller businesses or developing countries with limited resources.

2. Lack of awareness and education: Many individuals and communities may not fully understand the importance and benefits of sustainable development practices. Promoting awareness and providing education about the environmental, social, and economic advantages of sustainability is essential to encourage widespread adoption.

3. Policy and regulatory barriers: In some cases, existing policies and regulations may not support or incentivize sustainable practices. Governments and regulatory bodies need to develop frameworks and incentives that encourage sustainable development, including tax incentives, subsidies, or stricter regulations on unsustainable practices.

4. Limited access to technology and infrastructure: Sustainable practices often rely on advanced technologies, which may not be readily accessible or affordable in certain regions or sectors. Lack of infrastructure, such as renewable energy grids or waste management facilities, can also hinder the implementation of sustainable practices.

5. Short-term thinking and lack of long-term vision: Sustainable development requires a shift from short-term thinking and immediate gains to long-term planning and vision. This change can be challenging, as many decision-makers and stakeholders may prioritize immediate economic benefits over sustainability considerations.

6. Competing interests and stakeholder engagement: Sustainable development involves multiple stakeholders, each with their own interests and priorities. Balancing the needs of various groups, including businesses, communities, and environmental advocates, can be complex and may require extensive collaboration, negotiation, and compromise.

7. Cultural and behavioral barriers: Cultural norms, traditions, and societal behaviors can present barriers to the adoption of sustainable practices. Convincing people to change ingrained behaviors or practices

may require cultural sensitivity and finding alternative approaches that align with local values.

8. Measurement and reporting: Monitoring and measuring the impact of sustainable development practices can be challenging. Establishing reliable metrics and reporting frameworks to track progress and evaluate the effectiveness of sustainability initiatives is crucial to ensure accountability and drive continuous improvement.

Overcoming these challenges requires collective efforts from governments, businesses, communities, and individuals. It involves creating supportive policies, investing in research and development, promoting education, fostering collaboration, and addressing economic and social barriers to ensure a smooth transition towards sustainable development.

Individuals play a crucial role in promoting sustainable development. Here are some ways individuals can contribute to creating a more sustainable future:

1. Conserving resources: Individuals can actively conserve resources like energy and water by adopting simple habits such as turning off lights when not in use, using energy-efficient appliances, taking shorter showers, and reducing water wastage.

2. Reducing waste: Minimizing waste generation is important. Individuals can practice waste reduction by recycling, reusing items, composting organic waste, avoiding single-use plastics, and making conscious choices about consumption.

3. Sustainable transportation: Opting for sustainable transportation options like walking, cycling, using public transportation, or carpooling reduces carbon emissions and congestion. Individuals can also choose fuel-efficient or electric vehicles if possible.

4. Responsible consumption: Being mindful of consumption choices can have a significant impact. Individuals can support sustainable and ethical businesses, buy locally-produced goods, prioritize products with minimal packaging, and reduce unnecessary purchases.

5. Promoting renewable energy: Individuals can consider installing solar panels at home or supporting community-based renewable energy projects to promote the use of clean, renewable energy sources.

6. Advocacy and awareness: Individuals can raise awareness about sustainable development by engaging in conversations, sharing information with friends and family, and participating in community events. Advocacy can also involve supporting organizations and campaigns that promote sustainable practices.

7. Voting and engagement: Participating in elections and voicing opinions to policymakers can influence the development of sustainable policies. Individuals can support candidates or initiatives that prioritize sustainability, climate action, and environmental protection.

8. Education and learning: Continuously educating oneself about sustainability issues, attending workshops, and learning about sustainable practices can empower individuals to make informed choices and inspire others to do the same.

9. Supporting sustainable agriculture: Individuals can choose to support sustainable agriculture practices by purchasing organic, locally-produced, and ethically-sourced food. Additionally, growing one's own food through gardening can be a sustainable and rewarding practice.

10. Volunteerism and community involvement: Engaging in local sustainability initiatives, participating in clean-up events, volunteering for environmental organizations, or joining community gardens helps create positive change at a grassroots level.

Remember, even small actions can collectively make a significant impact. By adopting sustainable habits and encouraging others to do the same, individuals contribute to a more sustainable future for ourselves and generations to come.

Exploring renewable energy options for your home is a great way to promote sustainability and reduce your carbon footprint. Here are some common renewable energy options you can consider:

1. Solar power: Solar panels convert sunlight into electricity. Installing solar panels on your rooftop or in your backyard allows you to generate clean, renewable energy for your home. You can either purchase solar panels or lease them through power purchase agreements (PPAs). Additionally, you may be eligible for government incentives or tax credits to offset the initial cost.

2. Wind power: If you live in an area with steady winds, you can explore the possibility of installing a small wind turbine on your property. Wind turbines generate electricity as the wind turns the blades. However, it's important to assess factors like zoning restrictions, noise levels, and available space before considering this option.

3. Geothermal energy: Geothermal systems utilize the natural heat from the Earth's core to provide heating, cooling, and hot water for your home. A geothermal heat pump transfers heat between your home and the ground. While geothermal systems can be expensive to install, they offer long-term energy savings and minimal maintenance.

4. Biomass energy: Biomass energy utilizes organic materials such as wood pellets, agricultural waste, or dedicated energy crops to produce heat or electricity. You can consider installing a biomass boiler or stove to heat your home or a biomass power system to generate electricity. However, it's important to ensure a sustainable and responsibly sourced biomass supply to avoid negative environmental impacts.

5. Hydropower: If you have access to a flowing water source like a stream or river on your property, you can explore micro-hydropower systems. Hydropower turbines harness the kinetic energy of moving water to generate electricity. It's worth noting that the feasibility of hydropower depends on factors like water flow rate and the required permissions and permits.

Before deciding on a renewable energy option, consider the following steps:

1. Assess your energy needs: Evaluate your energy consumption patterns to determine the appropriate renewable energy system size for your home.

2. Conduct a feasibility study: Assess the available space, natural resources (sunlight, wind, etc.), and any local regulations or restrictions that may impact the installation of renewable energy systems.

3. Seek professional advice: Consult with renewable energy installers or experts to understand the costs, benefits, and technical considerations associated with each option.

4. Evaluate financial incentives: Research local, state, and federal incentives, tax credits, grants, or financing options that could help offset the initial investment costs.

5. Compare quotes: Obtain multiple quotes from reputable installers to compare pricing, warranties, system performance, and maintenance services.

6. Monitor and maintain: Once installed, monitor the performance of your renewable energy system regularly and follow the recommended maintenance guidelines to ensure its optimal operation.

Remember, each home is unique, and the feasibility of renewable energy options may vary depending on factors like location, budget, and available resources. Engaging with professionals and conducting thorough research will help you make an informed decision.

Solar power offers several benefits for homeowners. Here are some of the key advantages of using solar energy:

1. Renewable and Sustainable: Solar power is harnessed from the Sun, which is an abundant and renewable resource. As long as the Sun continues to shine, solar power will be available for us to harness and use.

2. Environmentally Friendly: Solar energy is a clean and green source of power. It does not produce greenhouse gas emissions or contribute to air pollution, making it an excellent alternative to fossil fuels. By switching to solar power, you can significantly reduce your carbon footprint and help combat climate change.

3. Energy Independence: Generating your own solar power allows you to become more self-sufficient and less reliant on traditional energy sources. This can provide a sense of energy independence, as you produce your electricity on-site, reducing your dependence on the grid and potential price fluctuations.

4. Cost Savings: While the initial investment for solar panel installation can be significant, solar power systems have the potential to save you money in the long run. By harnessing free sunlight, you can reduce your electricity bills and potentially even earn credits through net metering, where excess energy produced by your solar panels is fed back into the grid.

5. Long-Term Return on Investment: Solar panels have a long lifespan, typically ranging from 25 to 30 years or more. This means that once the initial cost of installation is recouped, you can enjoy many years of virtually free electricity. Additionally, solar energy systems generally require minimal maintenance, resulting in further cost savings over time.

6. Increasing Property Value: Homes with solar power systems tend to have higher property values and often sell faster than those without. Potential buyers recognize the benefits of solar energy and may be willing to pay a premium for a home with lower energy costs and a reduced environmental impact.

7. Job Creation and Economic Boost: The growing solar industry has led to the creation of numerous jobs in manufacturing, installation, maintenance, and research. By adopting solar power, you contribute to economic growth and job opportunities in the renewable energy sector.

It's important to note that the benefits of solar power can vary depending on factors such as location, system size, energy consumption, and local regulations. Assessing your individual circumstances and consulting with solar energy professionals will provide more accurate insights into the potential benefits for your specific situation.

Solar power offers several benefits, including:

1. Renewable and Clean Energy: Solar power relies on a renewable resource (the Sun) that is not depleted when used. It produces clean energy, without emitting greenhouse gases or other harmful pollutants, helping to reduce air pollution and combat climate change.

2. Energy Cost Savings: By using solar panels to generate electricity, you can save on energy costs. Solar power allows you to produce your own electricity, reducing your reliance on grid-supplied energy and potentially lowering your electricity bills.

3. Independence from Grid: Solar power promotes energy independence as it enables you to generate your electricity on-site. This can be beneficial during power outages or in remote areas with limited access to the grid.

4. Financial Incentives: Depending on your location, there might be financial incentives available for installing solar panels. These incentives can include tax credits, rebates, grants, or net metering programs, which allow you to sell excess electricity back to the grid. Taking advantage of these incentives can make solar power more affordable and provide a return on your investment.

5. Long Lifespan and Low Maintenance: Solar panels typically have a long lifespan (25 to 30 years or more) and require minimal maintenance. Once installed, they usually operate silently and have no moving parts, which reduces the need for costly repairs or frequent maintenance.

6. Job Creation and Economic Growth: The solar industry generates jobs and contributes to economic growth. As the demand for solar power increases, the industry creates employment opportunities in manufacturing, installation, maintenance, and research.

7. Environmental Benefits: In addition to reducing greenhouse gas emissions and air pollution, solar power also conserves water resources. Unlike traditional power generation methods, solar panels do not require significant amounts of water for operation.

It's important to consider factors like the suitability of your location, the size and efficiency of your solar system, local regulations, and your energy consumption patterns when assessing the benefits of solar power for your specific circumstances.

The financial incentives available for solar power can vary depending on your location. Here are some common incentives that you may find:

1. Solar Investment Tax Credit (ITC): In the United States, the federal government offers the Solar Investment Tax Credit, which provides a tax credit equal to a percentage of the cost of installing a solar energy system. As of 2021, the ITC offers a 26% credit for residential and commercial solar installations. However, please note that the percentage can change in future years.

2. Local and State Incentives: Many states, provinces, and local governments offer additional incentives to promote the adoption of residential and commercial solar power. These incentives may include rebates, grants, property tax incentives, or sales tax exemptions. Each jurisdiction has its programs and eligibility criteria, so it's worth researching what incentives are available in your specific area.

3. Renewable Energy Certificates (RECs): In some regions, you can earn Renewable Energy Certificates for generating solar power. RECs represent the environmental attributes of renewable energy generation and can be sold or traded to entities looking to offset their carbon footprint. The financial benefit of RECs can vary depending on market conditions and demand.

4. Net Metering: Net metering programs allow solar energy system owners to sell excess electricity they generate back to the grid, usually at the same price they would pay for electricity from the grid. This allows homeowners to offset their energy costs and potentially earn credits on their electricity bills.

5. Feed-in Tariffs: In certain locations, feed-in tariffs may be available. These programs offer long-term contracts where you are paid a fixed rate for the electricity your solar system generates. Feed-in tariffs can provide a guaranteed revenue stream and a quicker return on investment.

6. Solar Loans and Financing Options: Various financial institutions and organizations offer loans and financing options specifically for solar energy installations. These can include low-interest loans, green energy financing, or property-assessed clean energy (PACE) programs. Such financing options can help make solar power more affordable by spreading out the costs over time.

It's important to research and consult with local solar energy professionals or government agencies to understand the specific financial incentives available in your area. The availability, terms, and requirements of these incentives can vary, so it's essential to get accurate and up-to-date information for your situation.

B. Harnessing technology for ecological restoration

Harnessing technology for ecological restoration is an innovative approach that can help restore and rehabilitate damaged or degraded ecosystems. Technology can be used in various ways to enhance the efficiency, effectiveness, and scale of restoration efforts. Here are some examples of how technology is being utilized:

1. Remote Sensing and GIS: Remote sensing technologies, such as satellite imagery and aerial drones, combined with Geographic Information Systems (GIS), enable precise mapping, monitoring, and inventorying of ecosystems. These tools can provide detailed information about the location, extent, and condition of degraded areas, helping restoration practitioners plan and execute projects more effectively.

2. Data Analytics and Machine Learning: Advanced data analytics and machine learning algorithms can process large amounts of ecological data, providing insights into ecosystem patterns, dynamics, and potential restoration strategies. These technologies can help identify optimal sites for restoration, determine suitable species and planting patterns, and predict future changes in ecosystems.

3. Seed Banking and Biotechnology: Seed banks and genetic technologies play a crucial role in preserving and restoring plant species diversity. By storing the seeds of endangered or rare plant species, we can protect their genetic material for future restoration efforts. Biotechnology techniques, such as tissue culture and genetic modification, can also aid in propagating and enhancing plant species with specific traits for restoration purposes.

4. Hydrological Modeling and Restoration Planning: Hydrological modeling tools simulate water flow, infiltration, and distribution in ecosystems, helping restoration practitioners understand water dynamics and plan restoration projects accordingly. These tools can optimize the design of wetlands, stream restoration projects, and water management systems, ensuring that ecological processes are restored effectively.

5. Bioengineering and Eco-friendly Construction: Bioengineering techniques, such as the use of biodegradable materials and natural plant barriers, help control erosion and stabilize soil in restoration sites. Additionally, eco-friendly construction practices, such as permeable pavement and green infrastructure, can be applied to restore urban ecosystems and address stormwater management for improved water quality.

6. Citizen Science and Crowdsourcing: Technology facilitates citizen science initiatives where the public can actively participate in ecological restoration efforts. Smartphone apps and online platforms enable individuals to report observations, collect data, and contribute to monitoring and restoration projects. This collective effort can enhance the scale and accuracy of ecological restoration monitoring and research.

7. Environmental DNA (eDNA): eDNA refers to the genetic material organisms leave behind in their environment (such as water or soil). DNA sequencing technologies are used to identify species present in ecosystems, including rare or elusive species. eDNA analysis allows for a non-invasive and efficient method of monitoring biodiversity and can guide restoration efforts by identifying species composition and detecting the presence of target species.

By leveraging technology, ecological restoration efforts can be accelerated, refined, and adapted to achieve more successful and sustainable outcomes.

It is important to ensure that technology is used responsibly and in conjunction with ecological knowledge, local expertise, and community involvement to create the most effective restoration strategies.

Seed banks play a crucial role in ecological restoration by providing several benefits that contribute to the success and efficiency of restoration efforts. Here are some key benefits of seed banks:

1. Genetic Diversity Preservation: Seed banks preserve a diverse range of plant species and their genetic material, including indigenous, rare, and endangered species. This helps maintain biodiversity, which is essential for healthy ecosystems. Preserving genetic diversity allows for the reintroduction of locally adapted plant species, ensuring the resilience and long-term survival of restored ecosystems.

2. Availability of Native Plant Species: Native plant species are essential in ecological restoration as they are well-suited to the local environmental conditions and support the native wildlife and ecosystem functions. Seed banks collect, store, and distribute seeds of native plant species, making them readily available for restoration projects. This availability reduces the need to collect seeds from wild populations, which can impact their natural regeneration.

3. Restoration of Disturbed or Scarce Habitats: Seed banks often focus on collecting seeds from specific habitats that are degraded, rare, or under threat. By preserving seeds from these habitats, seed banks can support the restoration of these unique ecosystems. Restoring such habitats contributes to the conservation of specialized flora and fauna that depend on them.

4. Overcoming Seasonal and Geographic Limitations: Seeds collected from various plant species can be stored in seed banks for extended periods, overcoming seasonal restrictions and ensuring a continuous supply throughout the year. Furthermore, seeds in seed banks can be stored for long-distance transportation, allowing restoration projects to access plant materials from different regions, including remote or inaccessible areas.

5. Cost-effectiveness and Efficiency: Seed banks provide a cost-effective solution for ecological restoration projects. They eliminate the need for repetitive seed collection efforts, as seeds can be collected once and stored for future use. This reduces fieldwork costs, time-consuming seed collection endeavors, and the potential damage to existing native plant populations. Additionally, seeds in seed banks can be stored long-term, ensuring a resource for future restoration needs.

6. Disease and Climate Resilience: Seed banks can help mitigate the impact of diseases and climate change on plant species used in restoration projects. By preserving a wide range of genetic material, seed banks provide a gene pool that may include disease-resistant or climate-adapted traits. This allows the selection of seeds with desirable characteristics, enhancing the resilience of restored ecosystems.

7. Research and Adaptation: Seed banks serve as valuable resources for research and development. They provide opportunities for scientists to study plant genetics, conduct breeding experiments, and develop improved restoration techniques. Seed banks also support adaptive management approaches, allowing for the evaluation and adaptation of restoration strategies in response to changing environmental conditions or new knowledge.

In summary, seed banks are essential for ecological restoration efforts as they preserve genetic diversity, provide a continuous supply of native plant species, overcome seasonal and geographic limitations, offer cost-effectiveness and efficiency, enhance disease and climate resilience, and facilitate research and adaptation. By leveraging the benefits of seed banks, restoration practitioners can enhance the success and sustainability of their restoration projects.

While seed banks are valuable resources for ecological restoration, there are several challenges associated with seed banking that need to be addressed. Here are some key challenges:

1. Seed Viability and Dormancy: Seeds stored in seed banks have specific storage requirements to maintain their viability. Some seeds have short viability periods or specialized dormancy mechanisms, making long-term storage difficult. Ensuring the viability of stored seeds and breaking dormancy for successful germination can be challenging.

2. Genetic Integrity and Adaptation: Preserving the genetic integrity and adaptive potential of stored seeds is crucial for restoration success. However, the genetic diversity of a plant population can change over time in a seed bank due to genetic drift or unintentional selection. It is important to continually monitor and manage seed bank collections to maintain the desired genetic characteristics of the stored seeds.

3. Limited Seed Availability: Some plant species may have naturally limited seed production or face challenges in seed collection due to factors like small plant populations, dispersed or hard-to-reach habitats, or specific timing requirements. These limitations can make it difficult to gather an adequate quantity and diversity of seeds, affecting the availability of certain species for restoration projects.

4. Lack of Knowledge and Information: Comprehensive seed collection protocols, specific to each plant species, are required to ensure the quality and viability of stored seeds. However, collecting and storing seeds from many plant species still lack standardized protocols and detailed knowledge. Insufficient information on seed biology and ecology can hinder effective seed banking efforts and compromise the success of restoration projects.

5. Invasive Species and Contamination: Contamination of stored seeds by invasive plant species or pathogens can pose significant challenges. Ensuring that collected seed batches are free of invasive plants or contaminants is crucial to prevent unintentional introductions that could harm native ecosystems or compromise restoration goals.

6. Financial and Technological Constraints: Seed banking requires financial resources for seed collection, processing, storage facilities, and long-term maintenance. Setting up and maintaining seed banks can be costly, especially for large-scale or long-term projects. Additionally, technological advances are necessary to improve seed processing, preservation methods, and dormancy-breaking techniques.

7. Ethical and Social Considerations: Seed banking may raise ethical and social concerns related to intellectual property rights, access to genetic resources, cultural significance, and indigenous knowledge. Collaboration, inclusivity, and respect for local communities and their knowledge are essential to address these considerations and ensure the ethical use of seed resources.

Recognizing and addressing these challenges through research, collaboration, and investment can enhance the effectiveness and sustainability of seed banking for ecological restoration.

Standardizing seed collection protocols for different plant species requires careful consideration of their specific characteristics and needs. Here are some steps that can be taken to standardize seed collection protocols:

1. Research and Knowledge Gathering: Thoroughly study the biology, ecology, and reproductive characteristics of each plant species targeted for

seed collection. Identify specific seed traits, such as dispersal mechanisms, dormancy types, optimal timing for collection, and preferred storage conditions.

2. Develop Species-Specific Protocols: Based on the gathered knowledge, establish species-specific protocols detailing the best practices for collecting viable seeds. These protocols should include information on suitable collection methods, seed cleaning procedures, handling techniques to minimize damage or contamination, and appropriate storage conditions.

3. Consult Existing Resources and Guidelines: Consult existing literature, publications, and guidelines from botanical institutions, seed banks, and conservation organizations. Many resources provide general protocols for seed collection and handling that can be adapted to specific plant species.

4. Collaborate with Experts: Engage with botanists, ecologists, conservationists, and seed bank professionals who have expertise in the target plant species. Collaborative efforts can help refine and validate the collection protocols through their practical experience and knowledge.

5. Field Testing and Evaluation: Conduct field trials to test and evaluate the effectiveness of the seed collection protocols. Assess factors such as seed viability, germination rates, and genetic integrity to ensure that the protocols yield high-quality seeds suitable for ecological restoration projects.

6. Iterative Improvement: Continually review and update the seed collection protocols as new information and best practices emerge. Adapt the protocols to any changes in the understanding of the plant species' biology or advancements in seed collection techniques.

7. Standardization across Institutions: Encourage communication and collaboration among seed banks, botanical gardens, research institutions, and other relevant organizations. Seek consensus on best practices and promote the adoption of standardized seed collection protocols across different institutions to ensure consistency and comparability.

8. Knowledge Sharing and Education: Share the standardized seed collection protocols widely through training programs, workshops, publications, and online resources. Educating seed collectors, researchers, and restoration practitioners about the protocols and their importance will help build a collective understanding and promote adherence to standardized practices.

By following these steps and actively involving experts and stakeholders, it is possible to establish standardized seed collection protocols that enhance the effectiveness and comparability of seed banking efforts for different plant species.

C. Investing in renewable energy and regenerative practices

Investing in renewable energy and regenerative practices can have numerous positive impacts on the environment, economy, and society as a whole. Here are some key points to consider:

1. Renewable Energy: Investing in renewable energy sources such as solar, wind, hydro, and geothermal can help reduce dependence on fossil fuels, decrease greenhouse gas emissions, and mitigate climate change. Renewable energy technologies are becoming increasingly cost-competitive and have immense potential for long-term sustainability.

2. Economic Benefits: Investing in renewable energy can stimulate economic growth and job creation. The renewable energy sector offers opportunities for innovation, manufacturing, installation, maintenance, and research and development. It can also reduce energy costs for businesses and consumers over time.

3. Energy Security: Diversifying energy sources by investing in renewables decreases reliance on imported fossil fuels. This improves energy security, reduces geopolitical tensions, and increases resilience to price volatility and supply disruptions.

4. Environmental Sustainability: Renewable energy sources are typically cleaner and have lower carbon footprints compared to fossil fuels. They do not produce toxic air pollutants, contribute to acid rain, or generate hazardous waste. Investing in renewables helps protect air quality, water resources, and ecosystems.

5. Regenerative Practices: In addition to renewable energy, investing in regenerative practices in various sectors, such as agriculture and forestry, can help build resilience, restore ecosystems, and promote sustainable land and resource management. These practices include organic farming, agroforestry, reforestation, soil conservation, and sustainable water management.

6. Climate Change Mitigation: Renewable energy and regenerative practices play vital roles in mitigating climate change. By reducing greenhouse gas emissions and sequestering carbon, they contribute to achieving climate targets outlined in international agreements like the Paris Agreement.

7. Public Health Benefits: Shifting towards renewable energy sources can significantly improve public health by reducing air and water pollution associated with fossil fuel extraction and combustion. This can lead to fewer respiratory illnesses, improved air quality, and reduced healthcare costs.

8. Innovation and Technological Advancement: Investment in renewable energy and regenerative practices drives technological advancements and fosters innovation. Continued research and development can further improve the efficiency, scalability, and affordability of renewable energy technologies.

9. Social Equity: Investing in renewable energy can promote social equity by creating jobs and economic opportunities in communities that have historically been disproportionately affected by environmental pollution and energy poverty. This can lead to more inclusive and sustainable development.

10. Long-term Sustainability: Renewable energy and regenerative practices offer long-term solutions for meeting energy needs and managing natural resources sustainably. By investing in these practices, we can leave a healthier and more sustainable planet for future generations.

It's important to note that transitioning to renewable energy and regenerative practices requires a collaborative effort involving governments, businesses, communities, and individuals. Policy support, financial incentives, and public awareness are crucial for accelerating the adoption of renewable energy and regenerative practices worldwide.

Investments in renewable energy technology have the potential to revolutionize our energy systems and drive the transition towards a more sustainable and clean energy future. Here are some key points about investing in renewable energy technology:

1. Market Growth: The renewable energy sector has been experiencing significant growth in recent years, driven by advancements in technology, cost reductions, and increasing awareness about the need for clean energy. Investing in renewable energy technology provides opportunities to capitalize on this growing market.

2. Cost Competitiveness: The cost of renewable energy technologies, such as solar and wind, has been steadily declining, making them increasingly competitive with fossil fuel-based energy generation. This cost reduction is partly due to economies of scale, technological advancements, and increased manufacturing capacity. Investing in these technologies can offer attractive long-term returns through both capital appreciation and revenue generation.

3. Government Support: Many governments worldwide are actively promoting investments in renewable energy technology through various policy incentives. These include tax credits, grants, feed-in tariffs, and renewable energy targets. Government support can provide stability and a favorable investment environment for renewable energy technology.

4. Technological Advancements: Investments in renewable energy technology drive innovation and technological advancements. Research and development activities are focused on improving the efficiency, reliability, and scalability of renewable energy technologies. Investing in these advancements can lead to breakthroughs that further drive down costs, enhance performance, and expand the range of applications for renewable energy.

5. Energy Transition and Decarbonization: Investing in renewable energy technology is crucial for achieving energy transition goals and reducing carbon emissions. As countries and industries aim to decarbonize their energy systems, the demand for renewable energy technologies such as solar, wind, hydropower, and geothermal is expected to rise. Investing in these technologies positions investors for future growth in a low-carbon economy.

6. Job Creation and Economic Growth: Investments in renewable energy technology have the potential to create jobs and spur economic growth. The renewable energy sector offers employment opportunities across various stages of the value chain, including manufacturing, installation, operation, and maintenance. These investments can also attract new businesses and spur local economic development.

7. Sustainable Development: Investing in renewable energy technology aligns with the principles of sustainable development. By reducing

greenhouse gas emissions and promoting clean energy sources, these investments contribute to climate mitigation efforts, improve air quality, conserve resources, and mitigate environmental degradation.

8. Risk Diversification: Renewable energy technology investments can provide diversification benefits in investment portfolios. They often have a low correlation with traditional asset classes and can act as a hedge against the volatility of fossil fuel prices. Including renewables in an investment portfolio can help manage risk and enhance overall investment returns.

9. Social Impact: Investing in renewable energy technology offers social benefits, such as improved energy access and energy affordability. In regions where access to electricity is limited, investments in renewables, like off-grid solar systems, can provide clean and affordable energy solutions, empowering communities and improving quality of life.

10. Long-term Perspective: Investments in renewable energy technology have long-term potential. As the world transitions to a more sustainable energy mix, the demand for renewable energy is expected to grow significantly in the coming decades. Investing in renewable energy technology allows individuals and institutions to participate in this transition and position themselves for long-term financial and environmental benefits.

It's worth noting that, as with any investment, there are risks involved in investing in renewable energy technology. Factors such as policy changes, technological advancements, market dynamics, and project-specific risks should be carefully considered. Conducting thorough research and consulting with financial professionals can help investors make informed decisions.

Hydropower is a renewable energy source that harnesses the power of water to generate electricity. It has been a reliable and widely-used form of clean energy for many years. If you are interested in exploring investment opportunities in hydropower, here are some key points to consider:

1. Global Potential: Hydropower has significant global potential, with many countries relying on it as a source of renewable energy. Developing and emerging countries often have untapped hydropower resources, presenting opportunities for investment in new projects.

2. Renewable and Clean: Hydropower is a renewable energy source and does not emit greenhouse gases during electricity generation, making it a

clean alternative to fossil fuels. This characteristic contributes to sustainability and aligns with global climate goals.

3. Established Technology: Hydropower is a proven technology, with a long history of successful implementation. This established track record can provide investors with confidence regarding project stability and performance.

4. Scalability: Hydropower projects can range in size from small-scale installations to mega-dams. This scalability allows for flexibility in investment options, depending on your financial capacity and risk appetite.

5. Revenue Stability: Hydropower projects often benefit from stable and predictable revenue streams due to long-term power purchase agreements or feed-in tariffs. This stability can provide investors with consistent cash flows and enhance the overall attractiveness of hydropower investments.

6. Multiple Revenue Streams: Hydropower projects may generate revenue from multiple sources. In addition to electricity sales, ancillary services such as frequency regulation, capacity markets, and the potential for selling excess power back to the grid can contribute to project profitability.

7. Infrastructure Investment: Hydropower projects typically involve significant infrastructure investments, including dams, turbines, transmission lines, and power stations. These investments can create opportunities for companies specializing in construction, engineering, and equipment manufacturing.

8. Sustainable Development: Hydropower projects can have positive social and environmental impacts. They can contribute to regional development by creating employment opportunities, supporting local economic growth, and providing access to clean electricity in rural areas.

9. Project Risks: It's essential to understand the potential risks associated with hydropower investments. Factors like environmental impact, geological stability, regulatory changes, and community support should be carefully evaluated and mitigated to ensure long-term project viability.

10. Consider Regional Context: Different regions offer unique opportunities and challenges for hydropower investments. Factors such as water availability, government policies, grid infrastructure, and market dynamics should be examined when evaluating specific investment opportunities.

Before making any investment decisions, it's crucial to conduct thorough due diligence by assessing the feasibility, financial projections, environmental considerations, and regulatory landscape of potential hydropower projects. Engaging with industry experts, financial advisors, and experienced project developers can provide valuable insights and help make informed investment decisions in the hydropower sector.

Hydropower offers several environmental benefits, making it a desirable form of renewable energy. Here are some key environmental advantages of hydropower:

1. Greenhouse Gas Emissions Reduction: Hydropower generates electricity without burning fossil fuels, thereby reducing greenhouse gas emissions. This helps mitigate climate change and contributes to global efforts to reduce carbon footprints and combat air pollution.

2. Renewable and Sustainable Energy Source: Unlike non-renewable energy sources like coal, oil, or natural gas, hydropower relies on the continuous water cycle, making it a sustainable energy option. As long as there is water flow, hydropower can be harnessed to generate electricity without depleting finite resources.

3. Water Conservation: Hydropower does not consume water during electricity generation; it merely uses the energy of flowing water. While some water is temporarily impounded in reservoirs for power generation, it is eventually released downstream, allowing the water to continue its natural course. This helps maintain the ecological balance of water bodies and supports aquatic ecosystems.

4. Flood Control: Certain hydropower projects, such as dams, can contribute to flood control and management. By regulating water flow during periods of high rainfall, dams can mitigate the risk of flooding downstream, protecting communities and infrastructure.

5. Reservoir Benefits: The creation of reservoirs for hydropower generation can provide additional environmental benefits. Reservoirs often serve as recreational areas for boating, fishing, and other water-based activities. They can also act as habitats for wildlife, supporting biodiversity and enhancing ecosystem services.

6. Drought Resilience: In regions prone to drought, hydropower reservoirs can help mitigate the impacts of water scarcity. By regulating water release

during dry periods, hydropower can ensure a more reliable water supply for drinking, irrigation, and other essential needs.

7. Reliable and Predictable Energy Generation: Hydropower facilities provide a stable and consistent electricity supply, helping offset the intermittency of some other renewable energy sources like solar or wind power. This reliability supports grid stability and resilience.

However, it is important to note that hydropower projects may also have potential environmental challenges and impacts. These can include habitat alteration, sedimentation, changes in river flow patterns, and the displacement of local communities. It is crucial to carefully consider and address these potential impacts through appropriate project planning, environmental assessments, and mitigation measures to ensure sustainable and responsible hydropower development.

Overall, when properly managed and designed, hydropower can play a significant role in the transition to a low-carbon and sustainable energy future while minimizing adverse environmental impacts.

There are several alternatives to hydropower for generating renewable energy. Here are a few examples:

1. Solar Power: Solar energy harnesses the power of sunlight to generate electricity. Photovoltaic (PV) panels convert sunlight directly into electricity, while solar thermal systems concentrate sunlight to produce heat, which can then generate electricity. Solar power is abundant, widely available, and has minimal environmental impacts.

2. Wind Power: Wind turbines capture the kinetic energy from the wind and convert it into electricity. Wind power is a rapidly growing renewable energy source, with large-scale wind farms or smaller turbines installed on rooftops or in rural areas. It is clean, produces no emissions, and can generate power even in moderate wind conditions.

3. Geothermal Energy: Geothermal power harnesses the heat from the Earth's deep underground to generate electricity. This energy is obtained by tapping into hot water or steam reservoirs below the Earth's surface. Geothermal power is reliable, renewable, and has a smaller land footprint compared to other energy sources.

4. Biomass Energy: Biomass energy utilizes organic plant and animal materials to generate heat or electricity. Biomass can be in the form of

wood, agricultural residues, energy crops, or even organic waste. This renewable energy source can reduce the need for fossil fuel usage and waste disposal while providing a sustainable energy option.

5. Tidal Power: Tidal energy converts the kinetic energy of ocean tides into electricity using tidal turbines or barrages. As tides are predictable and consistent, this form of renewable energy has the advantage of being highly reliable. However, tidal power plants require suitable coastal locations and can have specific environmental considerations.

6. Wave Power: Wave energy devices capture the motion of ocean waves and convert it into electricity. They can be located offshore or near the shore. While still an emerging technology, wave power has substantial potential given the vast energy available in ocean waves.

7. Solar Thermal Energy: Solar thermal systems use mirrors or lenses to concentrate sunlight and produce heat, which can be used for electricity generation or other purposes like heating water or space. This form of renewable energy can be particularly effective in sunny regions.

These are just some examples of alternative renewable energy sources. Each has its advantages, challenges, and unique suitability in different geographic locations. The combination of multiple renewable energy sources is often the best approach to meet diverse energy needs, reduce reliance on fossil fuels, and mitigate environmental impacts.

Solar power and wind power are two popular forms of renewable energy that have seen significant growth in recent years. Here's a comparison of some key aspects between solar power and wind power:

1. Energy Source:
- Solar Power: Solar energy is derived from sunlight. Photovoltaic (PV) panels convert sunlight directly into electricity.
- Wind Power: Wind energy is generated by harnessing the kinetic energy from the wind using wind turbines, which then convert it into electricity.

2. Resource Availability:
- Solar Power: Solar energy is abundant and widely available. It can be harnessed as long as sunlight is present, although the intensity can vary depending on weather conditions and time of day.

- Wind Power: Wind energy requires a reliable and consistent source of wind to be effective. It is more readily available in certain geographical locations, such as coastal or open plains areas with high wind speeds.

3. Efficiency and Capacity Factor:
- Solar Power: Solar panels have a relatively high efficiency in converting sunlight into electricity, generally ranging from 15% to 25%. The capacity factor, which measures the actual electricity output compared to the maximum possible output, can vary depending on factors like location and panel orientation.
- Wind Power: Modern wind turbines typically have an efficiency of around 30-40% in converting wind energy into electricity. The capacity factor for wind power can also vary depending on wind conditions and turbine design, but it is generally higher than solar power, often ranging between 30% and 50% or more.

4. Land and Space Requirement:
- Solar Power: Solar power requires a larger land area for generating a significant amount of electricity. Large-scale solar farms or rooftops are commonly used. However, solar panels can also be installed on structures like parking canopies or incorporated into building materials.
- Wind Power: Wind power typically requires less land area per unit of electricity generated compared to solar power. Wind turbines can be erected in arrays or wind farms, and the surrounding land can often still be used for other purposes, such as agriculture or livestock grazing.

5. Environmental Impact:
- Solar Power: Solar power has minimal environmental impact. The operation of solar panels produces no air or water pollution, and there are no direct greenhouse gas emissions during electricity generation.
- Wind Power: Wind power is also considered environmentally friendly, as it produces no direct air or water pollution. However, wind turbines can pose challenges for birds and bats in terms of collision risks, and their visual impact may also be a consideration in some areas.

6. Cost:
- Solar Power: The cost of solar power has decreased significantly in recent years, making it more affordable and viable for many applications. Costs vary based on factors like system size, installation, and location.
- Wind Power: Wind power costs have also decreased, but they can still be higher compared to solar power. Costs depend on factors such as turbine size, site accessibility, wind speeds, and infrastructure requirements.

It's important to note that both solar power and wind power offer unique advantages and considerations. The choice between them often depends on factors such as resource availability, geographical location, energy demands, and specific project requirements. Incorporating a mix of renewable energy sources can provide a more diversified and reliable energy generation system.

Chapter 18 Parallel Paths to Cooperation

A. Exploring various case studies and scenarios of cooperative initiatives

In Chapter 18, "Parallel Paths to Cooperation," the focus is on exploring various case studies and scenarios of cooperative initiatives. This chapter delves into different examples of global cooperation and collaboration across various sectors and regions. Here are some key points covered in this chapter:

1. International Organizations and Agreements: The chapter examines how international organizations, such as the United Nations, World Trade Organization, and World Health Organization, facilitate cooperation among countries. It also explores the role of global agreements like the Paris Agreement on climate change and the Sustainable Development Goals in promoting collective action.

2. Cross-sectoral Cooperation: The chapter highlights the importance of cross-sectoral collaboration in addressing complex global challenges. It explores initiatives that bring together governments, businesses, civil society organizations, and academia to tackle issues such as poverty, health, education, and environmental sustainability.

3. Public-Private Partnerships: The chapter explores the role of public-private partnerships (PPPs) in promoting cooperation and achieving development goals. It discusses examples of successful PPPs in areas such as infrastructure development, technology transfer, and social entrepreneurship.

4. Regional Cooperation: The chapter examines regional cooperation efforts, focusing on different regional organizations and initiatives that promote collaboration among neighboring countries. Examples include the European Union, ASEAN, and the African Union, among others.

5. Case Studies: The chapter provides case studies that illustrate successful cooperative initiatives. These case studies cover a range of topics including peacebuilding, disaster response, sustainable agriculture, renewable energy, and sustainable tourism. Each case study highlights the challenges faced, the strategies employed, and the outcomes achieved through cooperation.

6. Challenges and Lessons Learned: The chapter also addresses the challenges and limitations of cooperation and draws lessons from past experiences. It explores factors that hinder cooperation, such as conflicting interests, lack of trust, and inadequate resources. It also discusses strategies to overcome these challenges and enhance the effectiveness of cooperative efforts.

Overall, Chapter 18 provides a comprehensive exploration of the diverse pathways to cooperation, emphasizing the importance of collaboration, dialogue, and collective action in addressing global issues and achieving sustainable development. It presents a range of case studies and scenarios to inspire readers and shed light on the potential of cooperative initiatives in shaping a better future for all.

Regional cooperation can have a significant impact on economic growth. When neighboring countries come together to foster cooperation and integration, it creates an environment that promotes trade, investment, and collaboration. Here are some key ways in which regional cooperation can positively affect economic growth:

1. Expanded Market Access: Regional cooperation often involves the establishment of preferential trade agreements, such as free trade areas or customs unions. These agreements eliminate or reduce trade barriers, such as tariffs or quotas, among member countries. By expanding market access, regional cooperation can boost exports, attract foreign direct investment, and stimulate economic activity.

2. Enhanced Business Environment: Cooperation among countries within a region can lead to harmonized regulations, standards, and procedures. This alignment can create a more predictable and business-friendly environment, reducing administrative burdens and regulatory barriers for businesses. A conducive business environment can attract investments, promote entrepreneurship, and encourage the growth of domestic industries.

3. Infrastructure Development: Regional cooperation often involves joint efforts to develop and improve infrastructure, such as transportation networks, energy grids, and telecommunications systems. Improved infrastructure connects markets, reduces logistics costs, and facilitates the movement of goods and services. This infrastructure development can enhance regional trade and connectivity, leading to improved productivity and efficiency.

4. Knowledge and Technology Transfer: Regional cooperation can facilitate the sharing of knowledge, expertise, and technology among member countries. This transfer of knowledge and technology can contribute to the development of local industries, innovation, and the adoption of best practices. It can also enhance the competitiveness of businesses and drive economic growth.

5. Collaboration in Specialized Sectors: Regional cooperation allows countries to pool resources and expertise in specialized sectors. For example, countries may collaborate in areas like research and development, healthcare, education, or tourism. By leveraging each other's strengths and capabilities, regional cooperation can lead to the development of competitive industries and the sharing of economic benefits.

6. Stability and Peace: Cooperation among neighboring countries promotes stability and peace in the region. When countries work together to address common challenges and resolve conflicts through peaceful means, it creates a conducive environment for economic growth. Political stability and security foster investor confidence, attract investments, and promote long-term economic development.

It is important to note that the impact of regional cooperation on economic growth can vary depending on the specific context, level of integration, and effectiveness of cooperation mechanisms. However, by fostering collaboration, integration, and the removal of trade barriers, regional cooperation has the potential to significantly contribute to economic growth in participating countries.

There are several examples of successful regional cooperation initiatives that have made a positive impact on economic growth. Here are a few notable examples:

1. European Union (EU): The EU is a prime example of successful regional cooperation. It began as a coal and steel trading agreement among six countries after World War II and has since evolved into a political and economic union of 27 member states. The EU has achieved significant economic integration through the free movement of goods, services, capital, and people. This cooperation has led to increased trade, investment, and economic growth among member states.

2. Association of Southeast Asian Nations (ASEAN): ASEAN is a regional organization comprising ten member states in Southeast Asia. ASEAN has worked towards building a single market and production base, promoting economic integration and cooperation in the region. Through initiatives such as the ASEAN Free Trade Area (AFTA), ASEAN has reduced tariffs and non-tariff barriers, facilitating trade and investment flows within the region. ASEAN has also benefited from collaboration in areas like tourism, infrastructure development, and disaster management.

3. West African Economic and Monetary Union (UEMOA): UEMOA is a regional organization consisting of eight West African countries that share a common currency, the CFA franc. UEMOA promotes economic integration and cooperation in the region by fostering a single market and monetary union. The common currency and harmonized regulations have facilitated trade and investment among member states, contributing to economic growth and stability.

4. Common Market for Eastern and Southern Africa (COMESA): COMESA is a regional economic community comprising twenty-one member states in Eastern and Southern Africa. COMESA promotes regional integration, trade, and collaboration among member countries. Through initiatives like the COMESA Free Trade Area, member states have removed trade barriers and implemented common customs procedures, leading to increased trade and investment flows in the region.

5. Pacific Alliance: The Pacific Alliance is a regional integration initiative formed by four Latin American countries: Chile, Colombia, Mexico, and Peru. The alliance aims to promote economic cooperation, investment, and trade among its members. It has implemented measures to facilitate the movement of goods, services, capital, and people within the region, enhancing economic integration and growth.

These successful regional cooperation initiatives demonstrate that when countries come together to foster collaboration, remove trade barriers, and promote economic integration, it can lead to increased trade flows, investment, and economic growth. These examples highlight the potential benefits of regional cooperation in driving sustainable economic development.

Regional cooperation initiatives can bring numerous benefits to participating countries. Some of the main advantages include:

1. Enhanced Trade Opportunities: Regional cooperation initiatives often involve the reduction of trade barriers such as tariffs, quotas, and customs procedures among member countries. By promoting regional trade integration, participating countries can access larger markets and increase export opportunities. This can lead to increased trade volumes, higher competitiveness, and improved economic growth.

2. Increased Foreign Direct Investment (FDI): Regional cooperation initiatives can attract higher levels of foreign direct investment. When countries collaborate and create a unified market, it becomes more attractive for foreign investors who seek to access a larger regional market instead of individual countries. The resulting FDI inflows can stimulate economic growth, create jobs, and boost technological advancements.

3. Economic Diversification: Regional cooperation initiatives often encourage economic diversification by promoting specialization and division of labor among member countries. By focusing on their comparative advantages, countries can optimize production capabilities and achieve economies of scale. This can lead to the development of new industries, increased productivity, and reduced dependency on a single sector.

4. Infrastructure Development: Regional cooperation initiatives often include joint projects and investments in infrastructure development such as transportation networks, energy grids, and telecommunications systems. These infrastructure improvements can enhance connectivity, facilitate the movement of goods and services, and reduce transportation costs. Improved infrastructure can also attract more investment, promote tourism, and unlock new economic opportunities.

5. Knowledge and Technology Transfer: Regional cooperation initiatives provide a platform for sharing knowledge, expertise, and technology among participating countries. This knowledge exchange can promote learning, innovation, and the adoption of best practices. It can also foster research and development collaborations, leading to advancements in various sectors and industries.

6. Political Stability and Peace: By fostering cooperative relationships, regional cooperation initiatives can help build trust and understanding among participating countries. This can contribute to political stability and reduce conflicts, creating a conducive environment for economic development and social progress.

Overall, regional cooperation initiatives offer participating countries the potential to access larger markets, attract investment, diversify their economies, improve infrastructure, and foster collaboration. These benefits can enhance economic growth, promote development, and improve the overall well-being of the countries involved.

Regional cooperation initiatives can promote peace in several ways:

1. Economic Interdependence: By fostering economic integration and cooperation, regional initiatives create interdependencies among participating countries. When nations rely on each other's economic well-being, they have a vested interest in maintaining peaceful relations to ensure the continued benefits of trade and investment. Economic interdependence can act as a deterrent to conflict, as it becomes more costly for countries to engage in hostile actions that could disrupt trade and cooperation.

2. Conflict Resolution Mechanisms: Regional cooperation initiatives often establish frameworks and mechanisms for resolving disputes peacefully. They provide platforms for dialogue, negotiation, and mediation, allowing participating countries to address conflicts, disagreements, and territorial disputes through peaceful means. These mechanisms can help de-escalate tensions, build trust, and prevent conflicts from escalating into larger-scale disputes.

3. Confidence-Building Measures: Regional cooperation initiatives often include confidence-building measures among participating countries. These measures can range from regular diplomatic meetings and information sharing to joint military exercises and security cooperation. By engaging in these activities, countries can enhance trust, mutual understanding, and transparency, reducing the likelihood of misunderstandings and conflicts.

4. Cultural Exchange and People-to-People Contacts: Regional initiatives often promote cultural exchange programs and facilitate people-to-people contacts among participating countries. By encouraging interaction and understanding between citizens of different nations, these initiatives can help break down stereotypes, prejudices, and misconceptions. Increased

cultural understanding can foster empathy, reduce tensions, and promote peaceful coexistence.

5. Cooperation in Non-Security Areas: Regional cooperation initiatives often extend beyond security concerns to various non-security areas such as trade, environment, health, and education. By collaborating on common challenges, countries can build trust, find shared interests, and work towards common goals. This cooperation establishes positive interactions and shared responsibilities, contributing to overall regional stability and peace.

6. Regional Integration as a Peacebuilding Strategy: Regional cooperation initiatives can be part of a broader peacebuilding strategy in regions affected by conflict or historical disputes. By bringing countries together under a common framework, these initiatives foster dialogue, address underlying causes of conflicts, and create opportunities for mutual cooperation and understanding. This process can contribute to long-term peace and stability.

While regional cooperation initiatives cannot guarantee peace on their own, they provide valuable platforms and tools to promote peaceful relations, prevent conflicts, and build trust among participating countries. Through economic interdependence, diplomatic mechanisms, confidence-building measures, cultural exchange, and collaboration in various areas, these initiatives play a significant role in promoting regional peace.

B. Engaging with diverse perspectives and alternative futures

Engaging with diverse perspectives and alternative futures is an essential aspect of regional cooperation initiatives. Here's why it is important:

1. Comprehensive Understanding: Engaging with diverse perspectives allows participating countries to gain a comprehensive understanding of various viewpoints, challenges, and opportunities. By considering different perspectives, regions can develop more inclusive and well-rounded strategies that take into account the diverse needs and aspirations of member countries. This leads to more effective and equitable regional cooperation.

2. Innovation and Creativity: Alternative futures thinking encourages thinking beyond the status quo and exploring innovative ideas and approaches. By embracing diverse perspectives and considering different possible scenarios, regional cooperation initiatives can foster creativity and

innovation. This can help identify new solutions to complex challenges, promote forward-thinking policies, and drive sustainable development in the region.

3. Resilience and Adaptability: Engaging with diverse perspectives and exploring alternative futures builds resilience and adaptability within regional cooperation initiatives. By considering a range of potential scenarios, initiatives can better prepare for different outcomes, anticipate challenges, and develop flexible strategies. This resilience enables regions to navigate unforeseen circumstances, adapt to changing circumstances, and bounce back from setbacks more effectively.

4. Inclusivity and Empowerment: Engaging with diverse perspectives ensures inclusivity within regional cooperation initiatives. It provides an opportunity for the voices of all member countries, regardless of size or economic power, to be heard and considered. This empowers smaller or marginalized nations to actively participate in decision-making processes, increasing their ownership and stake in the regional cooperation agenda.

5. Conflict Prevention: Engaging with diverse perspectives and exploring alternative futures can help prevent conflicts within regional cooperation initiatives. By considering different viewpoints and potential consequences, initiatives can identify and address underlying sources of tension or disagreements. This proactive approach to conflict prevention promotes dialogue, collaboration, and mutual understanding, reducing the likelihood of disputes escalating into conflicts.

6. Long-term Sustainability: Considering diverse perspectives and exploring alternative futures is crucial for long-term sustainability. It enables regions to identify and evaluate potential social, economic, and environmental impacts of their actions. By incorporating sustainability principles and considering different possible futures, regional cooperation initiatives can work towards achieving sustainable development goals and ensure the well-being of future generations.

In summary, engaging with diverse perspectives and exploring alternative futures is essential for fostering comprehensive understanding, promoting innovation and adaptability, ensuring inclusivity, preventing conflicts, and achieving long-term sustainability within regional cooperation initiatives. It helps build robust and effective strategies that address the diverse needs and challenges of the participating countries, leading to successful and impactful regional cooperation.

Exploring alternative futures offers several potential benefits. Here are some of them:

1. Anticipating Change: By exploring alternative futures, individuals and organizations can anticipate potential changes and trends that may occur. This allows for proactive planning and strategizing, helping to mitigate risks and exploit opportunities in advance. It enables individuals and organizations to stay ahead of the curve and adapt to changing circumstances more effectively.

2. Innovative Thinking: Exploring alternative futures encourages innovative thinking by challenging conventional assumptions and envisioning unconventional possibilities. It helps break free from the constraints of the present and inspires creativity in problem-solving. This can lead to the discovery of new ideas, solutions, and approaches that may have otherwise remained unexplored.

3. Strategic Decision-Making: By considering different alternative futures, decision-makers can make more informed and strategic choices. They can evaluate the potential consequences of different scenarios and assess the risks and benefits associated with each option. This helps in making decisions that are robust, adaptable, and aligned with long-term goals.

4. Adaptability and Resilience: Exploring alternative futures builds adaptability and resilience by fostering a mindset that is open to change. It prepares individuals and organizations to navigate uncertainty and unexpected events by developing the capacity to respond and adapt to different scenarios. This helps in maintaining relevance and competitiveness in a rapidly evolving world.

5. Reduced Surprises and Uncertainty: By exploring alternative futures, individuals and organizations can reduce the element of surprise and uncertainty. It helps in identifying weak signals and early indicators of potential disruptions or emerging trends. This allows for timely preparation, response, and mitigation strategies, minimizing the negative impacts of unexpected events.

6. Enhanced Vision and Strategy: Exploring alternative futures expands the vision and widens the perspective of individuals and organizations. It helps to develop a more comprehensive understanding of the complex systems, interconnections, and dynamics at play. This broader vision enables the creation of more robust and inclusive strategies that take into account the diverse possibilities of the future.

7. Increased Adaptation to Stakeholder Needs: By exploring alternative futures, organizations can better understand the diverse needs and expectations of their stakeholders. This helps in tailoring products, services, and initiatives to meet the changing demands and preferences of the target audience. It promotes customer-centricity and stakeholder engagement, increasing satisfaction and loyalty.

In summary, exploring alternative futures provides benefits such as anticipating change, fostering innovative thinking, facilitating strategic decision-making, building adaptability and resilience, reducing surprises and uncertainty, enhancing vision and strategy, and adapting to stakeholder needs. It is a valuable tool for individuals and organizations to navigate an ever-changing world and shape a successful and sustainable future.

Businesses can explore alternative futures by following these steps:

1. Environmental Scanning: Begin by conducting an environmental scan to identify trends, drivers, and signals of change that could impact the business. Look at technological advancements, social and cultural shifts, economic factors, political and regulatory changes, and industry trends. This will help in understanding the forces shaping the future landscape.

2. Scenario Planning: Develop multiple scenarios that represent different possible futures based on the identified trends and uncertainties. Each scenario should be plausible and represent a distinct set of circumstances. Consider a range of possibilities, including best-case, worst-case, and moderate scenarios.

3. Stakeholder Engagement: Engage with a diverse range of stakeholders, including employees, customers, suppliers, industry experts, and thought leaders. Seek their input, insights, and perspectives on the potential future developments. This helps in understanding different viewpoints, anticipating potential disruptions, and identifying emerging opportunities.

4. Collaborative Workshops: Organize collaborative workshops or brainstorming sessions with cross-functional teams within the organization. Encourage creative thinking and explore the implications of each scenario on the business. Identify potential challenges, risks, and opportunities associated with different futures.

5. Data Analysis and Modeling: Analyze available data and use modeling techniques to assess the potential impacts of each scenario on the business. Quantitative analysis helps in understanding the range of outcomes and the probabilities associated with each scenario. This enables businesses to make informed decisions based on a solid foundation of evidence.

6. Adaptive Strategies: Develop adaptive strategies that are flexible and can be adjusted based on the emerging developments in different scenarios. Consider how the business can leverage its strengths, exploit emerging opportunities, and navigate challenges in each potential future. Identify early warning signs and triggers for action in each scenario.

7. Continuous Monitoring and Review: Regularly monitor and review the external environment and key indicators to track changes and validate the assumptions made in the scenario planning process. Keep the scenarios updated and refine them as new information becomes available. This ensures that the business remains agile and responsive to evolving circumstances.

8. Learning and Iteration: As you navigate the future, learn from the outcomes and iterate your strategies accordingly. Adaptation and learning from both successes and failures are crucial for effectively exploring alternative futures and staying ahead of the curve.

By following these steps, businesses can effectively explore alternative futures, anticipate change, and develop adaptive strategies to thrive in an ever-changing marketplace.

Using data analytics for alternative future analysis can provide valuable insights for businesses. Here's how you can incorporate data analytics into the process:

1. Data Collection: Start by collecting relevant data from various sources, both internal and external to your organization. This can include historical sales data, customer behavior data, market research data, social media data, industry reports, and economic indicators. The more diverse and comprehensive your data sources, the better.

2. Data Cleaning and Preparation: Clean and preprocess the collected data to remove any errors, inconsistencies, or missing values. Transform the data into a structured format that can be easily analyzed. This step is essential to ensure the accuracy and reliability of your analysis.

3. Data Exploration and Visualization: Explore the data using descriptive analytics techniques. Analyze trends, patterns, and relationships within the data. Visualize the results through charts, graphs, and dashboards to gain a clear understanding of the current state and identify any emerging signals or outliers.

4. Predictive Modeling: Utilize predictive analytics techniques to build models that can forecast future outcomes based on historical data patterns. This can involve regression analysis, time series forecasting, machine learning algorithms, or other predictive modeling techniques. Use these models to simulate potential future scenarios and generate insights about potential outcomes.

5. Scenario Analysis: Apply your predictive models to different scenarios that represent alternative futures. Adjust the input variables and assumptions to reflect the context of each scenario. Analyze the predicted outcomes and assess their potential impact on the business. This can help you understand the risks, opportunities, and potential strategies needed for each potential future.

6. Sensitivity Analysis: Conduct sensitivity analysis by altering the key variables and assumptions in your models to see how the predicted outcomes change. This helps in understanding the drivers of change and quantifying the uncertainty associated with each scenario.

7. Decision-Making: Use the insights gained from data analytics to inform your strategic decision-making process. Evaluate the risks, benefits, and trade-offs associated with each alternative future. Develop adaptive strategies that can be adjusted as new data becomes available during the implementation phase.

8. Continuous Learning and Improvement: Monitor and assess the accuracy of your predictive models as you gain new data points and feedback from real-world outcomes. Continually refine your models, update your assumptions, and adjust your strategies based on the feedback loop of data analytics and continuous improvement.

By leveraging data analytics, businesses can enhance their understanding of alternative futures, make data-driven decisions, and increase their ability to navigate uncertainties effectively. Remember, while data analytics provides valuable insights, it should be complemented with other qualitative methods to develop a holistic understanding of the future landscape.

C. Demonstrating the transformative power of unity in different contexts

The transformative power of unity can be demonstrated in various contexts, including:

1. Social Movements: Unity among individuals who share a common cause can lead to significant social change. History has witnessed the success of numerous social movements, such as the Civil Rights Movement, Women's Suffrage Movement, and LGBTQ+ Rights Movement. Through unity, these movements have challenged societal norms, fought against discrimination, and brought about legislative changes.

2. Teamwork in the Workplace: Unity within a team or organization can create a transformative work environment. When individuals collaborate, share ideas, and support each other, they can achieve remarkable results. High-performing teams often exhibit unity in their shared goals, effective communication, and mutual trust. This unity can drive innovation, productivity, and overall organizational success.

3. Community Development: In communities, unity can play a vital role in bringing about positive change. When community members come together and unite around common goals like improving infrastructure, addressing social issues, or supporting educational initiatives, transformative development can occur. Unity empowers communities to advocate for their needs, enhance social cohesion, and create a sense of belonging.

4. Global Cooperation: The power of unity is also evident in the realm of global cooperation. International organizations, such as the United Nations, aim to foster unity among nations to address global challenges like poverty, climate change, and conflicts. Through collective action and collaboration, countries can work together to find solutions, promote peace, and create a better world for future generations.

5. Cultural Diversity and Inclusion: Unity can also transform societies by embracing and celebrating cultural diversity. When communities respect and value different cultural backgrounds, they create an environment that encourages learning, understanding, and acceptance. This unity promotes social harmony, reduces prejudices, and fosters multiculturalism, leading to a more inclusive and vibrant society.

In each of these contexts, the transformative power of unity demonstrates that when individuals come together, they can overcome obstacles, challenge established norms, and create positive change. Whether it's fighting for social justice, achieving shared organizational goals, or addressing global issues, unity is a catalyst for transformation in various aspects of life.

Unity can drive social change in several ways:

1. Collective Action: When individuals come together and unite around a common cause, they can amplify their voices and exert collective pressure on institutions and decision-makers. By mobilizing resources, organizing protests, signing petitions, and engaging in advocacy efforts, unified communities can bring attention to social issues and demand change.

2. Strength in Numbers: Unity provides strength in numbers, which increases the influence and impact of social movements. When large groups of people unite, they can create a powerful force that cannot be easily ignored. This unity can attract media attention, spark public discourse, and put pressure on those in power to address societal injustices.

3. Shared Resources and Expertise: Unity allows for the pooling of resources, skills, and expertise. By combining efforts, individuals and organizations can create a more comprehensive and effective approach towards addressing social issues. This can involve sharing financial resources, knowledge, networks, or even physical resources to maximize impact and create sustainable change.

4. Creating Awareness and Education: Unity helps in raising awareness about social issues by fostering dialogue, educating others, and challenging existing narratives. When individuals come together in solidarity and speak out about injustices, they can break the silence and bring attention to marginalized voices. By engaging in discussions, organizing workshops, or leveraging social media, unity can spread knowledge and foster a collective understanding of the need for change.

5. Building Alliances: Unity allows for the formation of alliances and collaborations between different groups or organizations working towards similar goals. By joining forces, diverse groups with different backgrounds and perspectives can create a broader and more inclusive movement for social change. Building alliances facilitates cross-sector cooperation, encourages mutual learning, and strengthens the impact of collective efforts.

6. Shifting Social Norms and Attitudes: Unity has the power to challenge and change societal norms and attitudes that perpetuate inequality, discrimination, or injustice. When individuals come together, they can challenge biased beliefs, break down stereotypes, and promote empathy and understanding. This unity can shift societal perceptions and foster a more inclusive and equitable environment.

7. Advocating for Policy Change: Unity plays a crucial role in advocating for policy changes at local, national, and international levels. By uniting around specific policy reforms, groups can lobby for legislative changes, engage in grassroots movements, and hold policymakers accountable. Unity can influence policy decisions that address systemic issues and promote social justice.

In essence, unity facilitates collective action, amplifies voices, and creates a strong and influential force for social change. Through coordinated efforts, shared resources, and a common vision, unity empowers individuals and communities to challenge the status quo, advocate for justice, and ultimately bring about systemic transformations in society.

There have been numerous examples throughout history where unity has successfully driven social change. Here are a few notable examples:

1. Civil Rights Movement (United States): The Civil Rights Movement in the United States during the 1950s and 1960s is a powerful example of unity driving social change. African Americans and their allies came together to challenge racial segregation and fight for equal rights. Through nonviolent protests, grassroots organizing, and mobilization, they successfully campaigned for landmark legislation like the Civil Rights Act of 1964 and the Voting Rights Act of 1965, paving the way for the advancement of civil rights.

2. Anti-Apartheid Movement (South Africa): The anti-apartheid movement in South Africa was a unified effort by various groups and individuals, both domestically and internationally, to challenge the oppressive system of racial segregation and discrimination. Led by figures like Nelson Mandela, the movement employed tactics such as boycotts, protests, divestment campaigns, and international pressure. The unity of diverse stakeholders contributed to the dismantling of apartheid and the eventual democratic transformation of South Africa.

3. Women's Suffrage Movement: The fight for women's suffrage, which sought to secure voting rights for women, gained momentum through unity and collective action. Women's suffrage movements were formed in different parts of the world, with activists organizing marches, protests, and advocacy campaigns. The unity among suffragettes and their persistence eventually led to significant victories, with women gaining the right to vote in various countries.

4. LGBTQ+ Rights Movement: The LGBTQ+ rights movement has achieved significant progress through unity and collective action. LGBTQ+ individuals and their allies have united to advocate for equal rights, non-discrimination, and the recognition of same-sex marriage. Through grassroots organizing, visibility campaigns, and legal challenges, the movement has successfully secured important legal protections and increased societal acceptance in many parts of the world.

5. Indian Independence Movement: The Indian independence movement, led by figures like Mahatma Gandhi, demonstrated the power of unity in achieving liberation from British colonial rule. Different groups and individuals united under the common goal of independence, employing tactics such as nonviolent resistance, civil disobedience, and nationwide boycotts. The unity of Indians across different social, cultural, and religious backgrounds played a crucial role in eventually gaining independence in 1947.

These examples highlight how unity can galvanize diverse communities and lead to transformative social change. From challenging racial discrimination and fighting for gender equality to advocating for LGBTQ+ rights and national liberation, these movements showcase the power of unified action in creating a more just and equitable society.

The Civil Rights Movement in the United States employed several key strategies to challenge racial segregation, discrimination, and fight for equal rights. Here are some of the strategies that were instrumental in the movement's success:

1. Nonviolent Resistance: A central philosophy of the Civil Rights Movement was nonviolent resistance, inspired by leaders such as Mahatma Gandhi. Activists, including Martin Luther King Jr., advocated for peaceful protests, sit-ins, boycotts, and marches to raise awareness and put pressure on institutions. Nonviolence was a powerful tool to expose the injustice of racial segregation, gain sympathy from the wider public, and maintain moral high ground.

2. Civil Disobedience: The movement employed civil disobedience, intentionally violating unjust laws to challenge the legal and moral legitimacy of segregation. Activists participated in sit-ins at segregated lunch counters, rode integrated buses during the Freedom Rides, and organized acts of civil disobedience that confronted discriminatory practices head-on.

3. Grassroots Organizing: The Civil Rights Movement took root in local communities, where grassroots organizations played a vital role. These organizations, such as the Southern Christian Leadership Conference (SCLC) and Student Nonviolent Coordinating Committee (SNCC), focused on organizing and mobilizing individuals at the local level. They conducted voter registration drives, established community centers, and organized meetings and workshops to empower individuals to take action against discrimination.

4. Legal Challenges: Another key strategy of the Civil Rights Movement was the pursuit of legal remedies to challenge discriminatory laws and practices. A significant example is the NAACP's strategy of strategically selecting cases to bring before the courts, culminating in landmark Supreme Court decisions such as Brown v. Board of Education, which declared racial segregation in schools unconstitutional. Legal challenges helped dismantle segregationist policies and establish legal precedents for equal rights.

5. Media Coverage: The Civil Rights Movement recognized the importance of media coverage in shaping public opinion and generating support. Activists strategically engaged with journalists and photographers to document the movement's actions and expose the realities of segregation and racial violence. Media coverage helped mobilize support, raise awareness, and create pressure for political change.

6. Coalition Building: The movement sought to build coalitions across different racial, ethnic, and religious groups. Activists understood the power of unity and collaboration, forging alliances with labor unions, religious organizations, and progressive white activists to create broader support for the cause. Coalition building helped amplify the movement's message and broaden its base of support.

7. Economic Boycotts: The Civil Rights Movement utilized economic boycotts as a means of exerting economic pressure and advancing their goals. For example, the Montgomery Bus Boycott in 1955-1956 protested the arrest of Rosa Parks and demonstrated the economic power of the

African American community. These boycotts aimed to hit discriminatory businesses' finances and compel them to desegregate.

These strategies, among others, helped galvanize the Civil Rights Movement, challenge systemic racism, and eventually lead to significant legislative victories, such as the Civil Rights Act of 1964 and the Voting Rights Act of 1965. The movement's strategies continue to inspire social justice movements globally, highlighting the power of organized grassroots efforts, nonviolent resistance, and the pursuit of legal remedies for transformative change.

Chapter 19 Epilogue: A New Dawn of Human Evolution

A. Reflecting on the journey from competition to cooperation

The journey of human evolution, as explored in "The AI-Integrated Human Evolution: From Competition to Cooperation," has indeed been an extraordinary one. As we reach the conclusion of this captivating volume, let us pause and reflect on the transformation from a world dominated by competition to one propelled by cooperation, leading us towards a new dawn of human existence.

Throughout history, competition has been deeply ingrained in our societies - from the struggle for limited resources to the pursuit of power and dominance. Yet, as we unveil the potential of futurism and transhumanism, we begin to glimpse a different path forward. A path that embraces collaboration, cooperation, and the power of collective intelligence.

In this epilogue, let us ponder the implications of this shift. What does it mean for our understanding of ourselves as individuals, as a species, and as part of a greater interconnected world? How can we harness the power of collaboration to overcome the challenges that lie ahead?

As we delve into the realms of transhumanism, we confront questions that push the boundaries of our imagination. Will our evolution lead us to a future where our biological limitations are transcended, where technology and humanity merge to unlock new capabilities and potentials? And if so, what ethical considerations must we grapple with along this transformative journey?

Moreover, the emergence of artificial intelligence brings about a complex landscape of possibilities. How can we ensure that AI remains a tool for progress and empowerment rather than a source of division and inequality? How can we navigate the intricate dynamics between humans and AI to create a symbiotic relationship that augments our collective capabilities?

As these inquiries unfold, it becomes clear that our journey towards transhumanism and the embrace of collaboration is not merely an intellectual exercise. It is a call for action, a call for us to actively shape the future we envision. Together, we have the opportunity to forge a path that transcends the limitations of the past, a path that celebrates diversity, inclusivity, and the collective pursuit of progress.

In this new dawn of human evolution, let us remember that our potential lies not only in the technological advancements we make but also in our ability to empathize, understand, and truly cooperate with one another. It is through collaboration that we enrich our perspectives, challenge our assumptions, and unlock new realms of creativity and innovation.

As we bid farewell to this volume, may it serve as a reminder that the future of transhumanism holds immense promise. It calls upon us to embrace cooperation, nurture our collective intelligence, and redefine what it means to be human.

Together, let us embark on this extraordinary journey, inspired by the boundless possibilities that await us.

B. Envisioning a harmonious future for humanity

As we envision the future of transhumanism and the integration of AI into our existence, let us strive for a future that is not only marked by technological advancements but also by harmony and balance.

In this harmonious future, the power of collaboration extends beyond mere cooperation. It transcends boundaries, bringing together diverse perspectives, cultures, and ideologies. It is through the recognition and embrace of our shared humanity that we can forge a future where unity and empathy prevail.

Imagine a world where technology serves as a tool for fostering connection and understanding. A world where AI-powered systems work hand in hand with humans to address pressing global challenges, such as climate change, poverty, and healthcare disparities. In this vision, transhumanism is not a disconnected endeavor but a unifying force that brings humanity together to champion a common cause.

In this future, societal structures are reimagined to prioritize the well-being and flourishing of all individuals. As AI and automation redefine the nature of work, we seek to ensure equitable distribution of resources, providing opportunities for personal growth, creativity, and fulfillment. We shift our mindset from competition to collaboration, recognizing that our collective intelligence is far greater than any individual's abilities alone.

Harmony also extends to our relationship with the natural world. We recognize our interconnectedness with the Earth and embrace sustainable practices that preserve and restore our environment. Technology and nature coexist in a synergistic manner, where advancements are made with a deep reverence for the delicate balance of ecosystems.

Philosophical questions arise as we consider the consequences and ethical implications of these advancements. How can we ensure that access to transformative technologies is equitable, bridging the gap between the privileged few and the marginalized? What safeguards and regulations must be in place to navigate the potential risks associated with AI integration? How do we address the concerns of privacy, autonomy, and human agency?

As we embark on this journey towards a harmonious future, let us navigate these questions with an open mind, humility, and a commitment to inclusive dialogue. Let us actively seek to understand the fears, aspirations, and concerns of all, fostering an environment of mutual respect and empathy.

Envisioning a harmonious future for humanity necessitates the recognition that our evolution transcends the technological realm. It is about nurturing our shared values, celebrating diversity, and cultivating a sense of collective responsibility. It is about embracing the power of collaboration to shape a world where every individual has the opportunity to thrive and contribute to the betterment of our species.

In closing, let us look forward with anticipation, not only to the future of transhumanism but also to the potential of a harmonious future where the integration of AI and human collaboration brings us closer to a more inclusive, compassionate, and interconnected world.

Fostering collaboration and unity among diverse perspectives is crucial for creating a harmonious future. Here are some ways we can work towards it:

1. Open and inclusive dialogue: Encourage open and honest conversations that embrace diverse viewpoints. Create spaces where individuals feel safe to express their opinions, ask questions, and challenge assumptions. Actively listen, seek to understand, and value different perspectives.

2. Empathy and understanding: Cultivate empathy by putting yourself in others' shoes. Seek to understand their experiences, values, and beliefs. Respect differences and be willing to learn from diverse perspectives. Bridge understanding by finding common ground and shared goals.

3. Education and awareness: Promote education and awareness about different cultures, backgrounds, and perspectives. Encourage learning about the histories, traditions, and contributions of different communities. Education helps break down stereotypes, fosters empathy, and enhances intercultural understanding.

4. Collaborative problem-solving: Engage in collaborative problem-solving to address shared challenges. Encourage diverse teams to work together, combining different strengths, skills, and perspectives. Emphasize the value of collective intelligence in finding innovative solutions.

5. Empowerment and inclusivity: Foster a sense of empowerment and inclusivity by giving a voice and decision-making power to individuals from diverse backgrounds. Create opportunities for meaningful participation, representation, and leadership. Value and respect the unique contributions each person brings.

6. Building bridges and networks: Facilitate connections between diverse communities, organizations, and individuals. Encourage cross-cultural exchanges, partnerships, and collaborations. Foster networks that promote dialogue, understanding, and cooperation.

7. Conflict resolution and mediation: Develop skills and processes for resolving conflicts and mediating disagreements. Encourage dialogue and negotiation to find common ground and reach solutions that respect diverse perspectives.

8. Promoting cultural exchange: Embrace cultural exchange programs, events, and initiatives that celebrate diversity. Encourage individuals to share their cultures, traditions, and experiences. These exchanges can foster mutual respect, appreciation, and understanding.

9. Advocacy and inclusiveness: Advocate for inclusivity, diversity, and equality in all aspects of society. Support policies, practices, and initiatives that promote fairness, representation, and social justice. Stand against discrimination, bias, and prejudice.

10. Continuous learning and growth: Recognize that fostering collaboration and unity is an ongoing process. Continuously learn, evolve, and adapt our understanding of diversity and inclusion. Be open to feedback and self-reflection to improve our own biases and preconceptions.

By actively engaging in these practices, we can foster collaboration and unity among diverse perspectives, creating a more inclusive, empathetic, and harmonious future for all.

Creating safe spaces for open dialogue is essential in fostering collaboration and unity among diverse perspectives. Here are some key steps to help establish such spaces:

1. Establish ground rules: Clearly define and communicate the expectations for respectful and inclusive dialogue. Encourage participants to listen actively, speak honestly, and refrain from personal attacks or judgment. Emphasize that everyone's opinion is valid and valued.

2. Ensure confidentiality: Promote an environment where participants feel comfortable sharing their thoughts and experiences. Assure confidentiality to build trust and encourage individuals to speak freely without fear of repercussions.

3. Provide active moderation: Assign a neutral moderator or facilitator who can guide the conversation, maintain a respectful atmosphere, and enforce the established ground rules. This person can step in if any conflicts arise and redirect the conversation towards productive and inclusive dialogue.

4. Active listening and empathy: Encourage participants to actively listen to one another, seeking to understand different perspectives and experiences. Emphasize the importance of empathy and create opportunities for individuals to share their stories, fostering a deeper understanding and connection.

5. Respectful language and tone: Encourage participants to use language and tone that promotes respect and inclusivity. Discourage offensive or derogatory language and remind individuals to use "I" statements to

express their opinions and experiences rather than speaking on behalf of others.

6. Encourage questions and curiosity: Create a space where questions are welcomed and curiosity is encouraged. Encourage participants to ask respectful questions to better understand different perspectives and challenge their own biases. This can spark meaningful conversations and promote learning.

7. Reflect on power dynamics: Be mindful of power dynamics that may be present within the dialogue. Encourage participants to be aware of their own privileges and actively work towards a more equitable and inclusive conversation. Ensure marginalized voices are given equal opportunities to be heard and respected.

8. Feedback and reflection: Foster a culture of continuous growth by encouraging participants to provide feedback on the dialogue process. Regularly reflect on the effectiveness of the safe space and make adjustments as necessary to create an environment that facilitates open, respectful, and productive conversations.

Remember, creating safe spaces for open dialogue requires a commitment from all participants to actively listen, respect, and learn from one another. These spaces can be instrumental in fostering collaboration, understanding, and unity among diverse perspectives.

Addressing conflicts during dialogue is crucial to maintaining a safe and productive environment. Here are some steps to help effectively address conflicts:

1. Recognize the conflict: As a facilitator or participant, be attentive to potential conflicts arising during the dialogue. Look for signs of tension, raised voices, or disrespectful language. Acknowledge that conflicts can be a natural part of a dialogue process and an opportunity for growth and understanding.

2. Pause and breathe: If tensions rise, allow for a moment of pause. Encourage participants to take a deep breath and collect their thoughts. This can help prevent an escalation of emotions and allow for a more constructive approach to resolving the conflict.

3. Active listening: Encourage the conflicting parties to actively listen to one another. Ensure each participant has an opportunity to express their

viewpoints and concerns without interruption. Emphasize the importance of seeking to understand and appreciate different perspectives.

4. Remain calm and respectful: Encourage participants to maintain a calm and respectful demeanor while addressing the conflict. Remind them to use "I" statements to express their feelings and avoid blaming or attacking others. Reinforce the ground rules for respectful dialogue.

5. Find common ground: Look for areas of agreement or shared values between the conflicting parties. Identifying common ground can help bridge differences and build a foundation for resolving the conflict collaboratively.

6. Mediate or seek assistance if needed: If the conflict becomes challenging to manage or resolve, it may be helpful to bring in a neutral third party to mediate or facilitate the dialogue. This person can help guide the conversation, ensure fairness, and provide additional perspective if necessary.

7. Practice empathy and perspective-taking: Encourage participants to put themselves in the shoes of others involved in the conflict. Encouraging empathy and perspective-taking can foster understanding and help de-escalate tensions.

8. Brainstorm solutions: Once both parties have had the opportunity to share their perspectives, encourage them to brainstorm potential solutions together. This collaborative approach can help find common ground and create win-win outcomes.

9. Focus on learning and growth: Encourage participants to view conflict as an opportunity for personal growth and learning. Remind them that conflict can deepen understanding, challenge assumptions, and lead to new insights.

10. Closure and reflection: After addressing the conflict, take a moment to reflect on what was learned and how the dialogue has progressed. Emphasize the importance of moving forward with respect and a commitment to ongoing dialogue and understanding.

Remember, conflict is a natural part of dialogue, and addressing it constructively can lead to deeper understanding and resolution. By fostering an environment of active listening, respect, and empathy, conflicts can be transformed into opportunities for growth and strengthened relationships.

C. Inspiring readers to embrace a cooperative mindset

To inspire readers to embrace a cooperative mindset, consider the following steps:

1. Highlight the benefits: Explain the advantages of adopting a cooperative mindset, such as improved communication, increased collaboration, and strengthened relationships. Emphasize how cooperation can lead to win-win outcomes and create a more harmonious and productive environment.

2. Share success stories: Share examples or anecdotes that demonstrate the positive impact of cooperation in various settings, whether it's in personal relationships, team projects, or global initiatives. Real-life success stories can inspire readers and show them the power of cooperation.

3. Discuss the value of diverse perspectives: Help readers understand that cooperation involves valuing different perspectives and recognizing the unique contributions that each individual brings. Encourage them to see diversity as an opportunity for growth, creativity, and innovation.

4. Empathy and understanding: Emphasize the importance of empathy and understanding in fostering a cooperative mindset. Encourage readers to put themselves in others' shoes, actively listen, and seek to understand different viewpoints. Highlight the benefits of open-mindedness and the willingness to compromise.

5. Collaboration over competition: Shift the focus from competition to collaboration. Explain that cooperation doesn't mean abandoning individual goals but rather finding ways to work together to achieve shared objectives. Illustrate how collaboration can lead to more sustainable and fulfilling outcomes.

6. Foster a sense of community: Encourage readers to cultivate a sense of community and interconnectedness. Explain how cooperation can create a supportive and inclusive environment in which everyone can thrive. Highlight the importance of building trust, respect, and cooperation within communities, whether it's at work, in schools, or within neighborhoods.

7. Provide practical strategies: Offer practical tips and techniques for readers to practice and implement a cooperative mindset in their daily

lives. These may include active listening, effective communication, conflict resolution, and seeking win-win solutions. By providing actionable steps, readers can feel empowered to embrace cooperation.

8. Promote cooperation in various domains: Illustrate how cooperation applies to different areas of life, such as family dynamics, friendships, work environments, and even global issues like climate change or social justice. Show readers how a cooperative mindset can create positive change at different scales.

9. Facilitate dialogue and collaboration: Encourage readers to actively engage in dialogue and collaborative efforts with others. Suggest ways to initiate conversations, build consensus, and work towards shared goals. Provide resources or platforms that support cooperation and collaboration, whether it's through community organizations, online communities, or local initiatives.

10. Celebrate cooperation: Finally, celebrate and acknowledge instances of cooperation and collaboration. Highlight examples and achievements that demonstrate the power of working together. By recognizing and appreciating cooperative efforts, readers will be further motivated to embrace a cooperative mindset.

Remember, inspiring readers to embrace a cooperative mindset requires emphasizing the benefits, sharing relatable stories, fostering empathy, providing practical strategies, and showcasing the positive impact of cooperation across various domains.

Cooperation plays a vital role in addressing global challenges. Here are some ways in which cooperation can help tackle these issues:

1. Shared resources and knowledge: Global challenges like climate change, poverty, and public health require collective action and the sharing of resources and knowledge. Cooperation allows countries and organizations to collaborate, pool their resources, and share expertise for developing innovative solutions and implementing effective strategies.

2. Collective problem-solving: Global challenges are often complex and multidimensional, requiring diverse perspectives and expertise. Cooperation facilitates collective problem-solving, as different countries and stakeholders can contribute their unique insights and experiences. By combining efforts and working together, solutions can be developed that address the diverse needs of different nations and communities.

3. Strengthening global governance: Many global challenges require strong global governance mechanisms to coordinate and implement actions effectively. Cooperation can help strengthen international institutions, agreements, and collaborations, ensuring that there is a shared commitment to addressing these challenges. Multilateral cooperation promotes coordination, accountability, and the enforcement of global regulations and standards.

4. Resource mobilization: Cooperation enables countries to leverage their combined resources and mobilize support for addressing global challenges. Through partnerships and collaborative initiatives, countries can share the financial burden, access funding, and pool resources to implement large-scale projects. Such cooperation allows for more significant impact and improves resource allocation for addressing global challenges.

5. Knowledge sharing and capacity building: Cooperation promotes the sharing of best practices, research, and expertise among countries and organizations. Through collaborative efforts, knowledge and skills can be transferred, capacity can be built, and lessons learned can be shared. This allows for a more efficient and effective response to global challenges, as countries can learn from each other and implement proven strategies accordingly.

6. Building collective resilience: Global challenges often transcend borders and affect multiple regions or countries simultaneously. Cooperation helps build collective resilience by fostering trust, solidarity, and mutual support. By working together, countries can strengthen their capacity to prevent and respond to crises, such as natural disasters, disease outbreaks, or economic downturns.

7. Addressing inequalities: Cooperation can also help address the global challenge of inequality. By working together, countries can promote fair trade, reduce poverty, enhance access to education and healthcare, and address social injustices. Cooperation ensures that resources and opportunities are distributed more equitably and that no one is left behind in the pursuit of sustainable development.

Overall, cooperation is essential in addressing global challenges as it allows for shared resources, collective problem-solving, strengthening global governance, resource mobilization, knowledge sharing, capacity building, building collective resilience, and reducing inequalities. By fostering cooperation at local, regional, and international levels, we can work towards a more sustainable, equitable, and prosperous future for all.

International collaborations bring numerous benefits and hold great importance in today's interconnected world. Here are some key reasons why international collaborations are important:

1. Shared knowledge and expertise: International collaborations allow for the exchange of knowledge, expertise, and research findings across borders. By working with experts from different countries and cultures, new perspectives and innovative ideas can emerge, leading to advancements in various fields such as science, technology, medicine, and environmental research.

2. Addressing global challenges: Many of the challenges we face today, such as climate change, pandemics, poverty, and terrorism, are global in nature and require collaborative efforts to find effective solutions. International collaborations enable countries to pool their resources, share best practices, and coordinate actions, enhancing the collective response to these challenges.

3. Cultural understanding and diversity: Collaboration between individuals and organizations from different countries fosters cultural understanding and respect. It encourages the appreciation of diverse perspectives, beliefs, and traditions, promoting tolerance and reducing misunderstandings or biases. By embracing cultural diversity, international collaborations contribute to building more inclusive and harmonious societies.

4. Economic growth and development: International collaborations can lead to economic growth and development for participating countries. By partnering with other nations, countries can access larger markets, attract foreign direct investment, and boost trade opportunities. Collaborative projects and research initiatives can also drive innovation, create jobs, and stimulate economic advancement.

5. Peace and diplomatic relationships: Collaborative efforts can contribute to fostering peaceful relationships and resolving conflicts between nations. Through joint initiatives, dialogue, and exchanges, international collaborations encourage cooperation and understanding, reducing tensions and promoting a more peaceful and stable world.

6. Capacity building and infrastructure development: International collaborations often involve sharing resources, expertise, and technologies. This can lead to capacity building in areas such as education, healthcare, infrastructure development, and governance. Developing countries can benefit from the knowledge transfer and support provided through international collaborations, allowing them to improve their social and economic systems.

7. Scientific and technological advancements: Collaborations between scientists, researchers, and institutions from different countries drive scientific and technological advancements. Joint research projects enable sharing of data, resources, and expertise, accelerating discoveries and breakthroughs in various fields. These advancements have far-reaching implications, improving quality of life, addressing global challenges, and promoting sustainable development.

8. Cultural, educational, and artistic exchange: International collaborations in the cultural, educational, and artistic spheres promote cross-cultural understanding and appreciation. Collaborative projects, student exchanges, and artistic collaborations contribute to the enrichment of societies, broadening horizons, and fostering creativity and innovation.

International collaborations offer immense potential for addressing global challenges, promoting cultural understanding, driving economic growth, and advancing scientific breakthroughs. By working together across borders, we can harness the collective power of diverse perspectives, talents, and resources to create a better world for everyone.

International collaborations play a crucial role in addressing climate change by fostering global cooperation, knowledge sharing, and coordinated action. Here are some ways in which international collaborations contribute to tackling climate change:

1. Sharing best practices and knowledge: International collaborations provide a platform for countries to share their experiences, strategies, and best practices for mitigating and adapting to climate change. By learning from each other's successes and failures, countries can develop more effective policies and actions to reduce greenhouse gas emissions, conserve resources, and build resilience.

2. Coordinated action and targets: Through collaborations like international agreements (e.g., Paris Agreement), countries set common goals, targets, and timelines for reducing emissions and transitioning to low-carbon economies. By collectively working towards these goals, countries can ensure that efforts are aligned, avoiding duplication and enhancing effectiveness.

3. Technology transfer and capacity building: International collaborations facilitate the transfer of climate-friendly technologies and knowledge from developed to developing countries. Developing nations can benefit from the technical expertise, financial support, and capacity building programs provided by their counterparts, enabling them to implement sustainable solutions and transition towards cleaner energy sources.

4. Financial support and investment: Collaborative initiatives encourage developed countries to provide financial support and investments to developing nations to undertake climate change mitigation and adaptation measures. International funds, such as the Green Climate Fund, help in financing projects that promote clean energy, sustainable agriculture, and climate resilience in vulnerable regions.

5. Collective monitoring and reporting: International collaborations establish mechanisms for monitoring and reporting progress on climate change actions. Regular reporting and transparency help to hold countries accountable for their commitments, identify gaps, and track global progress towards emission reduction targets.

6. Research and innovation: Collaborative research projects between scientists and institutions from different countries advance the understanding of climate change impacts, mitigation strategies, and adaptation measures. Sharing research findings and collaborating on innovative solutions leads to the development of new technologies, practices, and policies to combat climate change effectively.

7. Global advocacy and awareness: International collaborations amplify the collective voice and influence of participating countries in advocating for climate action on the global stage. By collectively advocating for stronger commitments, policies, and funding, countries can drive greater ambition and urgency in addressing climate change.

8. Synergistic efforts and problem-solving: Climate change is a global problem that requires the collective effort of all nations. International collaborations encourage countries to work together, pooling resources, expertise, and experiences to find holistic and integrated solutions. By leveraging the diverse strengths and perspectives of different countries,

collaborations can yield innovative approaches to mitigating climate change and achieving sustainable development.

Overall, international collaborations provide a platform for countries to come together, share knowledge, and take coordinated action to address climate change. By fostering a spirit of cooperation and solidarity, collaborations enhance the effectiveness and impact of individual efforts, driving global progress towards a more sustainable and climate-resilient future.

Final Words and Conclusion:

As we reach the final word of this exhilarating journey, "The AI-Integrated Human Evolution: From Competition to Cooperation," we stand at the precipice of an awe-inspiring future. We have delved into the possibilities of futurism, witnessed the transformative potential of transhumanism, and explored the remarkable power of collaboration. Together, we have embarked on a captivating adventure that shapes the very essence of what it means to be human.

Throughout these pages, we have discovered that the path forward lies not in the isolated pursuit of individual victories, but in the embrace of collective synergy. We have uncovered the immense power that arises when we shed the shackles of competition and instead embark on a journey of cooperation. It is through unity and collaboration that we unlock the full potential of our collective intelligence, pushing the boundaries of what is possible.

Dear readers, we extend our heartfelt gratitude to you for accompanying us on this voyage of discovery. By investing your time and curiosity, you have played an invaluable role in shaping the narrative of our shared future. As we part ways, we invite you to carry the flame of hope and anticipation in your hearts.

For the future that awaits us is one brimming with exciting possibilities—a fusion of human and artificial intelligence, where cooperation propels us towards greatness. A future where borders blur, and global collaborations flourish. A future where the collective wisdom, innovation, and empathy of humanity intertwine harmoniously with the transformative potential of technology.

Indeed, the journey ahead is not without challenges, but it is through collaboration that we will find the resilience to triumph over adversity. By harnessing the power of AI and joining hands across borders, cultures, and disciplines, we can overcome the barriers that lie before us and build a world that thrives on cooperation, compassion, and sustainable progress.

So, let us embrace this future with open arms and open minds. Let us continue to explore, innovate, and collaborate, for it is in our unity that we find strength. Together, we hold the power to shape a future where the remarkable potential of humanity is fully realized.

Thank you, dear readers, for your trust and companionship on this remarkable journey. Let us step forward into the horizon of possibilities, instilled with the excitement and determination to forge a better world. The future beckons, and we are poised to answer its call.